The Art of Learning

The Art of Learning

Neural Networks and Education

Francis T.S. Yu, Edward H. Yu, and Ann G. Yu

CRC Press is an imprint of the
Taylor & Francis Group, an informa business

CRC Press
Taylor & Francis Group
6000 Broken Sound Parkway NW, Suite 300
Boca Raton, FL 33487-2742

CRC Press is an imprint of Taylor & Francis Group, an Informa business

Printed on acid-free paper

International Standard Book Number-13: 978-0-8153-6129-9 (Hardback)

Library of Congress Cataloging-in-Publication Data

Names: Yu, Francis T. S., 1932- author. | Yu, Edward H., author. | Yu, Ann G., author.
Title: The art of learning : neural networks and education /
Francis T.S. Yu, Edward H. Yu and Ann G. Yu.
Description: Boca Raton, FL : CRC Press, an imprint of the Taylor & Francis Group, [2019] |
Includes bibliographical references.
Identifiers: LCCN 2018021685 (print) | LCCN 2018032474 (ebook) |
ISBN 9781351116343 (Master eBook) | ISBN 9781351116336 (Adobe PDF) |
ISBN 9781351116329 (ePUB) | ISBN 9781351116312 (Mobipocket) |
ISBN 9780815361299 (hardback)
Subjects: LCSH: Learning–Physiological aspects. | Cognitive learning. |
Paired-association learning.
Classification: LCC QP408 (ebook) | LCC QP408 .Y8 2019 (print) | DDC 612.8/2–dc23
LC record available at https://lccn.loc.gov/2018021685

Typeset in Minion
by Integra Software Services Pvt. Ltd.

Visit the Taylor & Francis Web site at
http://www.taylorandfrancis.com

and the CRC Press Web site at
http://www.crcpress.com

To All Learners

I was NOT supposed to . . . but I DID
I was NOT supposed to graduate
from high school, but I DID
I was NOT encouraged
to go to college, but I DID
I could NOT afford to go to the United
States,
but I DID
NO teacher ever thought I'd likely be
admitted to
The University of Michigan,
but I WAS
I was NOT expected to get a Ph.D.,
but I DID
I did NOT think I could be a professor at a
major American university,
but I DID
I was NOT supposed to publish any books
or technical papers,
but I DID (over 300!)
I did NOT expect to be known, let alone
to be outstanding in my field,
but I AM
I was NOT supposed to become a profes-
sor and scholar,
but I DID
I WAS SUPPOSED to be an excellent
athlete
. . . ***but I wasn't***
-F.T.S. Yu

Contents

Preface

Though I have been a scientist and university professor for almost my entire adult life, I didn't always show an aptitude for scholarship. In fact, students and colleagues who know of my academic accomplishments are often amazed to discover that I was once considered by both teachers and peers to be a troublemaker, delinquent, and overall dummy. They are often further surprised (and sometimes appalled) to know that I barely made it through high school due to my penchant for talking back to teachers, or simply ignoring them, shunning homework, and cutting classes. (Why stay in school when you can put your feet up at the local theater to enjoy the air-conditioning and catchup on the latest Hollywood release?) In fact, the disparity between who I appeared to be and what I ended up achieving has created confusion in the minds of many over the years, leading them to question how I could have succeeded despite all signs initially pointing to failure. This disparity has led many to ask me about my "secrets" to becoming successful.

In truth, I harbor no secrets and relatively few thoughts on the issue of success. However, I do possess many thoughts about the phenomenon of learning and why it is much more important to clarify, and thereby understand, what it actually is rather than to focus narrowly on getting good grades, beating out one's classmates, moving up the academic ladder, acquiring a respected title, getting rich, or otherwise attaining the trappings of what our dominant culture deems success. As a result, I would like to redirect people's attention to what I consider to be a more fundamental and important question: What does it mean to learn?

By now, you may be thinking that this is not your typical textbook. You may even be sensing (correctly) that it isn't even a textbook at all. In fact, I'm not sure where it belongs—perhaps an unnamed category in the bookstore which is one part pedagogy, two parts philosophy, three parts science, not to mention a smidgeon of electrical engineering, and a dose of epistemology. Mainly, I have written this book because I believe what is happening in schools and universities often has less to do with actual learning and more with operant conditioning. To be more specific, mass education in this day and age is increasingly about programming people's minds in a machine-like fashion, and this has the unfortunate effect of deadening, not only human intelligence, but human sensibilities. It is consequently my hope that by better understanding human learning, and thereby differentiating it from the conventional and dehumanizing norm, we will begin to reclaim our intelligence, our sensibilities and, in turn, our shared humanity.

–by Edward Yu (as told by Francis Yu)

CHAPTER 1

Common Sense Training

Let me start with some common sense in education. In Figure 1, we see an Asian water buffalo and in Figure 2 an Arabian horse. When we try to project for the apparent strengths and

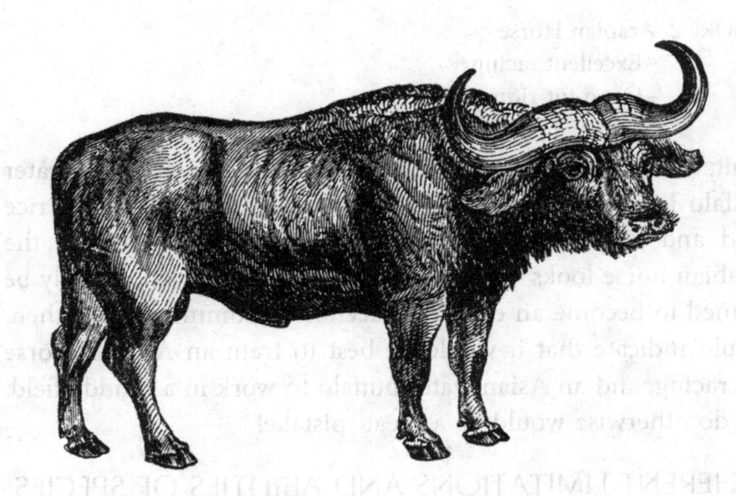

FIGURE 1 Asian Water Buffalo
–Excellent for working in a muddy rice field
–Good swimmer

FIGURE 2 Arabian Horse
–Excellent racing
–Good for riding

limitations between the two in pictures, we see that the water buffalo looks to be very well suited for work in a muddy rice field and might also be a very good swimmer, whereas, the Arabian horse looks very good for riding and could probably be trained to become an excellent racehorse. Common sense, then, would indicate that it would be best to train an Arabian horse for racing, and an Asian water buffalo to work in a muddy field. To do otherwise would be a great mistake!

INHERENT LIMITATIONS AND ABILITIES OF SPECIES

Is one better than the other? Common sense tells us we must understand the inherent advantages of each one.

CHAPTER 2

Audio vs. Visual Learning

I have always struggled with language. While I was born in a foreign land and English is my second language, I seem to have a more difficult time with it than my wife, who is also not a native English speaker. She excels at picking up the nuances of the spoken word, is a talented writer, can be detail oriented and has the amazing ability to repeat conversations verbatim. I, on the other hand, often appear not to be listening closely. I like to think visually, instead of verbally and enjoy figuring out how things work. Where language is concerned we are different: she is an Arabian horse and I am a water buffalo (or vice versa!).

Individuals vary in the way they process information. One key difference in learning styles is sequential vs. spatial learning.

Sequential learners are able to listen skillfully and navigate verbal instructions quite easily, learning in a step-like manner. Often they are great storytellers and can mimic sounds and conversations easily. Due to this auditory skill, they are known as **Auditory-Sequential Learners**. These people excel at all things verbally related and are often talented writers.

Spatial learners are visually attuned, often excelling at pattern recognition and puzzling out the overall picture. Due to their visual acuity, they are known as **Visual-Spatial Learners**.

There are advantages to both learning styles and most people possess a little of each quality while skewing more towards one or the other. Unfortunately, in an effort to streamline the learning process, education has historically seemed to favor the Auditory-Sequential Learners. Why? A.S. learners are good listeners and are able to navigate oral instructions quite aptly. As a result, they are "easier" to teach. School tends to be a positive experience for them. They are able to express themselves well, are good listeners, learn sequentially, are rapid processors, and think verbally.[1]

Visual-Spatial Learners learn via imagery of the whole concept, are spatially aware, astute observers, think holistically, and need more time to process information.[2]

It can sometimes take V.S. learners longer to express what they are learning and, as a result, they may sometimes seem to be lacking. For them, early academic life is often a negative experience. At least, as a youngster, it was that way for me. In fact, it wasn't until my last year in high school that I started to recognize my strengths.

Fortunately, the situation is improving. Today there is a better recognition of different learning styles (Figure 3), including of those on the autism spectrum.

I would also like to introduce a third style of learning called the **Narrative Learner**. The Narrative Learner is someone who requires a story to make sense of our world. That story could be told in images or in words and—because of its sophisticated nature—often requires both. Since a story is usually expressed in words, this is where Auditory-Sequential Learners can excel as they have an ear for the sounds and cadences of language. Also of importance in this style is its sequential nature. Time is an important factor in its expression.

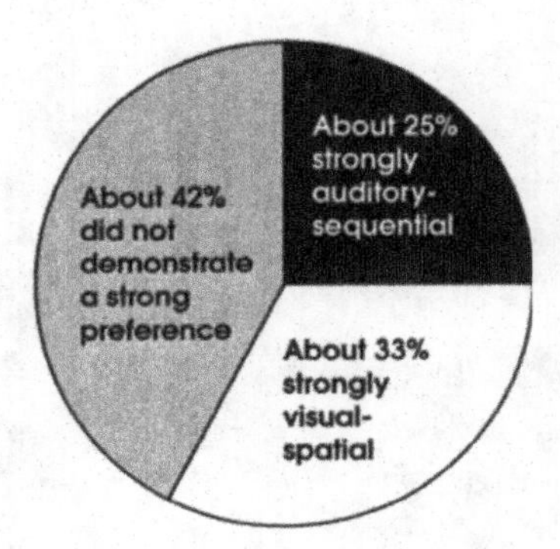

FIGURE 3 Distribution of students with strongly preferred learning styles in the regular classroom.

Note. From the visual-spatial resource website (http:/www.visualspatial.org). Adapted with permission.

All this is to say that we must recognize the inherent strengths and weaknesses of all types of learning styles whether it be auditory-sequential, visual-spatial, or narrative. Remember, the water buffalo is not an Arabian horse!

NOTES

1 Linda Kreger Silverman, Ph.D. *Identifying Visual-Spatial and Auditory-Sequential Learners: A Validation Study.*

2 Ibid.

CHAPTER 3

Human Brains vs. Machines

As I hinted earlier, a fundamental problem with our educational system lies in the narrow focus it makes on transferring a massive amount of information from an authority figure (namely the teacher) to a subject (namely the student), while granting little to no encouragement, or even time, for critical examination of the information. In practice, the problem often lies in making students memorize large amounts of information in a relatively short period of time, without regard to any understanding of where the information might have come from (i.e. who procured it and whose interests might be served by its dissemination), why exactly they should memorize it, what remains unknown and unexplored, or simply gets overlooked or ignored in the dissemination process, and how they themselves might make their own discoveries and come to their own conclusions. In short, a conventional education often treats human beings more like mindless machines than sensing and feeling organisms capable of thinking, wondering, reasoning, creating, and imagining.

Obviously, we are different from machines, and to ignore this fact is not only to render education inefficient in the cultivation of critical thought, reasoning, creativity, and imagination, it is, in a sense, a misuse or even abuse of our humanity.

If we force humans to perform machine-like tasks, we not only risk being far less efficient, accurate, and precise in accomplishing those tasks, but, more importantly, risk dehumanizing the human being. Conversely, if we expect even the most sophisticated machines to behave like humans we are entering the world of science fiction or fantasy. Talk to any competent scientist or engineer who works in robotics, pattern recognition, and artificial neural networks, for example, and she will tell you that even the most high-tech devices currently available cannot come remotely close to matching people in performing our most basic functions. These differences are why, even today, most robots can only perform the most rudimentary and unvarying movements compared to humans. They are able to walk and talk in a primitive fashion, but only within a minimal amount of variation and degree of freedom.

Now let us look at the major differences between neural networks and digital computers. A digital computer has at least one, and sometimes several, central processing units, while a neural network has many (i.e. neurons). Digital computers use sequential processing, while neural networks use parallel processing. Digital computers use programs to execute, while neural networks use associative learning rules.

Therefore, digital computers are very good at computational processing while neural networks are good at cognitive operations such as pattern recognition, understanding a language, etc.

In terms of calculations such as addition, subtraction, multiplication, and division, a computer is more efficient than a human brain

But for cognitive types of operations such as understanding a language, pattern recognition, or thinking, a three-year-old child, or even a dog, can perform better than that of the best supercomputer (Figure 4)!

FIGURE 4 Human's "best friend"

The survival of early humans was very much dependent upon their ability to run, climb, and reproduce, as well as their cognitive ability to hunt in groups, seek shelter, and use tools. It took their brains millions of years to evolve from an early primitive state to our current neural network. In this process, the parts they often used expanded, while those of little use died away or shrank.

If conditions had been different and more dependent on calculation, then our brains would be different too and might behave more like a computer.

But because they weren't, the human brain is plastic and malleable and good for cognitive learning but has a poor storage capacity and is slow in processing compared to a computer.

This is why it is a mistake to force human beings to try and behave like computers. If we know this, and embrace it, it will be to our advantage.

For cognitive types of operations, such as understanding a language or creative thinking, a three-year-old child, or even a dog, can perform better than that of the best supercomputer!

BRAIN PLASTICITY

Our brains consists of billions of neural networks that allow us to learn and adapt. Figure 5 shows an illustration of a human brain made up of neural networks. Neural networks are massively interconnected via synapses. Through cloning, scientists have been able to reproduce and grow a human brain, but even with the help of this reproduction these networks remain mysterious.

To produce from scratch a facsimile of neural networks, we must necessarily understand some of the concrete ways the brain works. That is why, in some of my research as a scientist, I have endeavored to create artificial neural networks. Figure 5 includes my representation of one.

Techniques to simulate our unique human neural networks are basically drawn from cognitive psychology and from biological models. The purpose of such research is to simulate the structure of a network of massively interconnected biological neurons in which information is processed in a parallel as well as an associative manner.

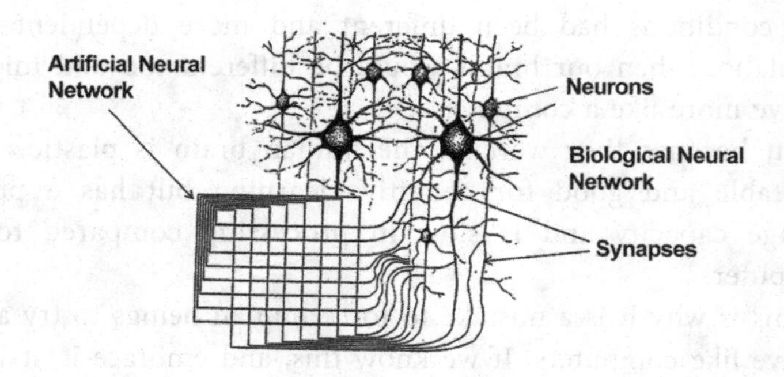

FIGURE 5 Biological and artifical neural networks

One thing I would like to point out is that artificial neural networks (ANN) do NOT work the same as a computer. There are fundamental differences between the two. In ANN, every single neuron has the ability to perform simple logical operations. In contrast, for digital computing *every* computation must go back to, and rely on, a central processing unit (CPU). Note that in the past few years quantum computing has made advances, but is still currently in its infancy.

Moreover, ANNs are capable of learning, improving their performance, adapting to different environments, and coping with disruptions, whereas, digital computers are programmed with rigid rules that must be manually reprogrammed for better solutions.

Neural networks can process inexact, ambiguous, fuzzy data that do not exactly match stored information.

Electronic computers cannot adjust the specific definitions and rules for which they are programmed to accommodate new, inexact, degraded, or contradictory input.

Each neuron has many input signals, but only one output signal, which is fanned out to many pathways that are connected to other neurons. These pathways interconnect with other neurons to form a network, as illustrated in Figure 6. They can be also **self-interconnected.**

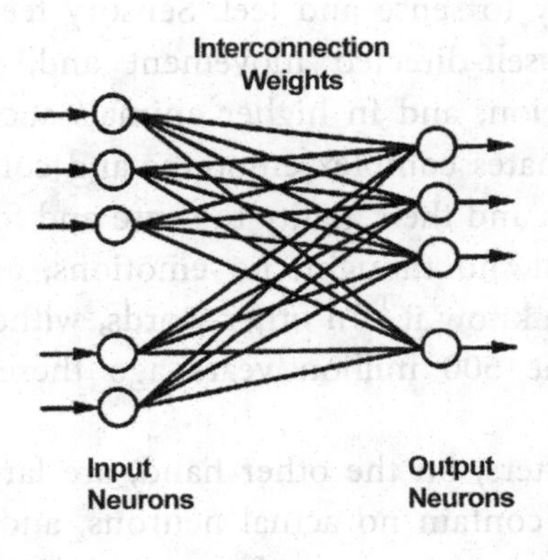

FIGURE 6 A one-layer artificial NN

The operation of a neuron is determined by a transfer function that defines the neuron's output as a function of the input signals. Every connection entering a neuron has an adaptive coefficient, called weight, assigned to it. The weight determines the interconnection strength between neurons, and it can be changed by a learning rule that modifies the weights in response to the input signals and the value supplied by the transfer function. The learning rule allows the response of the neuron to change with time, depending on the nature of the input signals. This means that the network adapts to its environment and organizes information within itself, resulting in a learned response.

Still, there are advantages to digital computing which is why we use computers in our daily lives.

The key, however, is to USE them as tools to supplement our intelligent and very creative human brains. As of this writing, computers are not meant to operate as replacements for them.

The Neuron

Humans possess a nervous system made up of billions of neurons, each one of which connects to upwards of ten thousand of its near to distant neighbors. Neurons provide, among other important things, sensory feedback, or to put it in simpler terms, our ability to sense and feel. Sensory feedback, in turn, makes possible self-directed movement and, therefore, self-directed exploration, and in higher animals such as elephants, whales, and primates complex emotions and complex thinking. Without neurons and their ability to sense and feel, there would be no movement, no thought, no emotions, and, indeed, no animal life as we know it.[1] In other words, without the birth of the neuron some 500 million years ago there would be no animal kingdom.

Digital computers, on the other hand, are largely comprised of silicon chips, contain no actual neurons, and, consequently, cannot sense and feel. Some machines, such as jet airliners and robots, possess sensors whose electronic feedback gets fed into computer algorithms, all of which, in turn, help to direct the machine's movements. The sensory feedback in all machines, however, is highly primitive compared to the types available to people. All responses, furthermore, are generated by algorithms that have been programmed by humans.

It is for this reason that digital computers are superior at following algorithms and predetermined patterns but inferior at responding appropriately to novel information, or to generate innovative responses to familiar information, adapt to new circumstances, and respond to unfamiliar environments. Digital computers are, for the most part, incapable of learning and are only capable, in general, of following human-designed algorithms and responding to input in a stereotypical fashion. A software designer may change a computer's algorithms, but this constitutes learning on behalf of the software designer and not the computer.

Finally, computers can only process information in a *sequential* fashion, receiving it strictly in the order that it is given, bit by literal bit. In contrast to a computer's sequential processing of information, people are able to "chunk" or "compress" huge swaths of information. This makes it possible for them to skip steps, fill in blanks, make cognitive leaps, draw inferences, correct for inconsistencies, and multitask. In engineering terms, this means we are able to process information in a *parallel* fashion.

The general inability of digital computers, and the vast majority of other machines, to learn this way leads to enormous deficits on their part, which is why, at this time, they can only complement the human brain and not replace it when it comes to recognizing patterns, coordinating movements, communicating, and performing other quintessentially human functions. These functions are not just human traits, but common to all mammals and, indeed, virtually all members of the animal kingdom. An animal's abilities rely crucially on these functions provided by its neurons. No machine yet is made of anything quite like them.[2]

Differences between Thinking and Computing

A computer is capable of making computations and storing information through digital switching devices. Its operations are strictly electronic, and, before the digital age, mechanical and initiated by the human user (though in some cases monkeys, parrots, and other members of the animal kingdom could probably stand in). It is the user's thinking, sensing, feeling, and emoting that determine what operations are set into motion and what information gets stored into the machine's memory. The computer itself has no ability to think, sense, feel, or emote and therefore no way to make any decisions that reside outside a preprogrammed algorithm. In other words, something analogous to a decision making ability might be coded into a computer's software, but it is rooted in fixed algorithms and, therefore, not facilitated by any organic and variable process.

Human beings can also compute and store information, but only in the unique and minimally understood fashion that neurons make possible. Though it has become commonplace and even customary to borrow electronics terms such as "hardware," "hardwire," "software," "programming," and "circuitry" when referring to the human mind, it is important to remember that human physiology contains no electronic components. Failing to realize this very basic and important fact can lead us to using such terms literally rather than figuratively and therefore fool ourselves into believing that the human mind actually functions like a digital computer.

Conversely, digital computers cannot think, imagine, reason, or learn. To equate their capacities with ours is not only inaccurate but highly demeaning. Why is this important? Because if we take what a computer does, which is mainly to compute figures and store information in a highly rigid and unvarying fashion, and call this intelligent behavior, we are not only degrading our intelligence but possibly severely limiting it as well. Thinking, imagining, reasoning, and learning will always reside strictly within the domain of human capabilities.[3] Digital computation and storage will always remain within the exclusive domain of machines.[4]

PROCESS VS. PRODUCT

One of the most important challenges of life is to adopt lifelong learning methods. The question should always be are we learning effectively? Current neural science has shown us that our brains are quite malleable. Genetically speaking we have been born with excellent neural networks. In a nurturing educational environment, I believe that it is possible for an average youngster to develop into a very intelligent and successful adult.

But what is a nuturing educational environment? Can it be found in an atmopshere of rigorous discipline and exposure to large quantities of material?

Or is it in a world that fosters creative processes that lead to creative thinking? My belief is that it is the latter.

For us to become smarter, we need to foster a creative process-oriented environment that leads to personal curiosity and development. In order to achieve this goal, we need to know the inherent abilities and limitations of the human mind. Human beings have shown themselves to be, time and time again, infinitely creative and curious. This means that innovative ways of teaching and learning are essential.

NOTES

1 It can also be argued that plants can sense and feel. The difference is that because they lack neurons, they have to do so in a manner qualitatively different from that of animals.

2 This includes artificial neural networks. An artificial neural networks is itself, merely a crude simulation of its biological counterpart.

3 All members of the animal kingdom are capable of learning. Members of species that possess a more complex nervous system are also capable of some form of thinking, imagining and reasoning.
4 Like the digital computers of today, humans have, in the past, been employed to perform long and often tedious calculations. In fact, humans were employed to perform such tasks starting a couple centuries before the machines that eventually took over their efforts were even invented. This is why before the digital age, the term *computer* was commonly used in reference to the people who were paid to make the needed calculations or otherwise follow the algorithms that would provide the desired results. Nonetheless, in this book, unless otherwise specified, I use the conventional definition of computer to refer to the machine and not its human predecessor.

CHAPTER 4

Simplicity in Learning

> Knowledge is a process of piling up facts; wisdom lies in their simplification.
>
> Martin H. Fischer

One definition of complexity is the ensemble of groups of smaller simple systems. An automobile is assembled by thousands of more simple mechanical and electrical parts (Figure 7).

A human body is actually a very large-scale biologically integrated (VSBI) system. It is a daunting task to learn the detailed construction of either system, unless you are actually working on it. The fact is that our brains are not designed to remember or equipped to learn very complicated systems or complicated equations. Even my own books or technical papers, after a couple of years of not using them, I cannot remember most of the details! If we know our limitations it is usually to our advantage. It is much more difficult for our brains to learn complex systems or complicated equations. However it is easier (and *more enjoyable*) to learn the basics and the fundamental concepts.

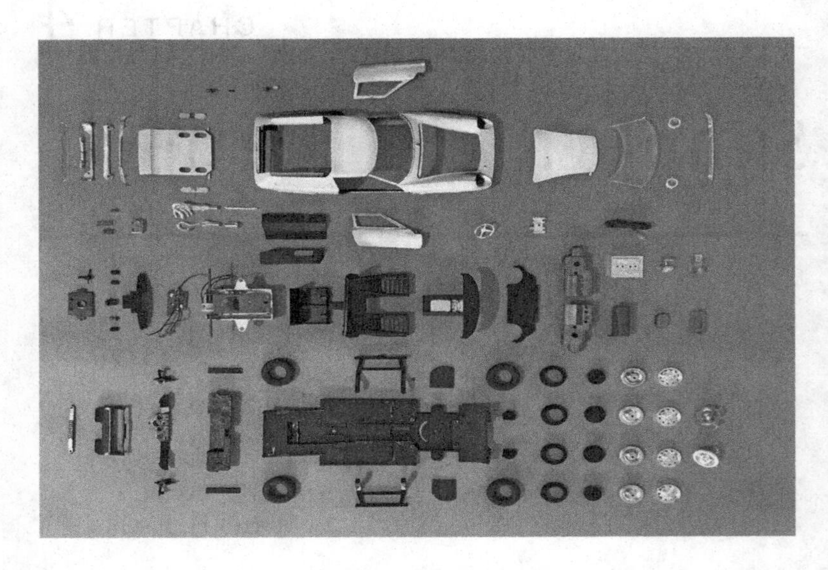

FIGURE 7 This photo of a vintage 1:20 scale model Lotus 47GT was inspired by an original photo created by Volkswagen in 1988. The original photo was a VW Golf, disassembled into its 6,843 parts. (The 1988 photo was created by Hans Hansen as part of a book made by Volkswagen for its employees on its 50th anniversary. Commissioned by Volkswagen Co, Art Director Dietmar Meyer.)

> An automobile is assembled by thousands of more simple mechanical and electrical parts

Endeavoring to find simplicity in learning is especially beneficial during the **process** of learning. Maintaining simplicity in learning brings with it a sense of accomplishment and success while motivating an individual to continue on to more complex issues. This process nurtures one's natural curiosity, fundamental to further self-training and unsupervised learning.

The human brain has billions and billions of neurons which are massively interconnected with billions and billions of synapses. Does this mean our brains are purposely designed to remember billions and billions of bits of information? Absolutely not! In fact it is a serious mistake to ask our neural network to overburden its capability. In our experience, it is more efficient to ask our neural network to understand simple components and put them together later to build a more complex integrated system.

Note that the human brain is not designed to remember a huge quantity of information, although we have billions and billions of neurons! Otherwise, we would not have developed those storage gadgets such as computers, compact discs (CDs), flash memory sticks, and others to help us.

Learning subjects such as mathematics and science becomes much easier when we break up complex ideas into much smaller pieces. This is not anything new, but again, teachers and students alike are much too eager to jump into the important work of calculating equations vs. understanding concepts. I must stress, the important work is actually understanding the basics and fundamentals. This is far more important than calculations and the memorization of complex equations. In the following sections, I would like to demonstrate a few examples of learning simply.

> If you can't explain it to a six-year-old, you don't understand it well enough
>
> Albert Einstein

MAXWELL'S EQUATIONS

Learning Simply: Assemblage of Parts

I have often posed this question: Other than Einstein and Newton, what scientist's laws are the most widely applied in science? The answer would be James Clerk Maxwell, who, in the 1860s first assembled a set of four differential equations that form the theoretical basis for describing classical electromagnetism. This has come to be known as **Maxwell's Equations** and includes

- Gaus's law
- Gaus's law for magnetism
- Faraday's law
- Ampère's law.

By assembling all four of Maxwell's equations together and providing the correction to Ampère's law, Maxwell was able to show that electromagnetic fields could propagate as traveling waves. In other words, Maxwell's equations could be combined to form a wave equation. Maxwell's insight stands as one of the great theoretical triumphs of physics.

The fact that Maxwell never came up with his own **original** equations does not negate their combined significance. In fact, it was his brilliance in recognizing the combination of these four laws that led to the creation of a hugely powerful tool. Indeed, Maxwell was the first to provide a theoretical explanation of a classical electromagnetic wave and, in doing so, *compute* the speed of light. Thus without Maxwell's equations, Einstein would have been less likely to discover the theory of relativity!

My point here is to make an illustration of how an assemblage of simple parts can create a much more powerful whole (and the vice versa of a more complicated whole being able to be break down into more simple parts).

LEARNING CALCULUS

Learning Simply: The Concept behind the Equation

What is Calculus? Calculus most likely started as a basic tool for calculating areas of irregular curved shapes. One can clearly see, for instance, that if one wishes to estimate the area "A" of an irregular shape of farmland (Figures 8 and 9), one would divide the farmland into many narrow rectangular shapes. For simplicity we assumed they are all equal in width, as shown on the left-hand side in the slide. Then it is apparent that the estimated

Calculus can be seen as an extension from geometry, using a new set of symbolic representations, illustrated below:

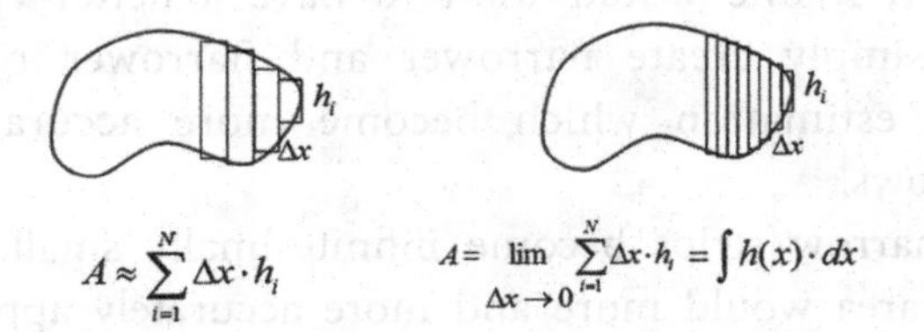

h_i
Δx
h_i
Δx
$A \approx \sum_{i=1}^{N} \Delta x \cdot h_i$
$A = \lim_{\Delta x \to 0} \sum_{i=1}^{N} \Delta x \cdot h_i = \int h(x) \cdot dx$

The symbolic representation (or transformation, or even substitution as you want to call it) makes calculus more efficient to apply. Note that the symbolic representation actually to puts calculus in a higher level of mathematical abstraction.

FIGURE 8 The art of learning calculus

FIGURE 9 An irregularly shaped rice field. The ancients used a form of calculus to calculate curved irregularly shaped areas.

area can be easily calculated as the addition of all those areas, as given by the equation on the left-hand side in Figure 8.

However, if one would want to have a better estimation, one can simply create narrower and narrower rectangular strips for estimation which become more accurate as the strip narrows.

If the narrow strips become infinitesimally small, then the estimated area would more and more accurately approach the actual area of the farmland.

Now as we let the width of the rectangles approach zero and replace the summation sign by an integral sign. By incorporating the use of the summation sign we then have so-called **Newtonian calculus.** (In reality, calculus was actually developed by Leibniz.)

Let me emphasize that, by employing **symbolic representation** in calculus we are able to make it more efficient to apply. Notice that this symbolic representation further abstracts the mathematical equation: The symbology puts calculus at a higher level of abstraction of mathematical operation!

I would like to stress that the usage of symbolic representation in mathematics is very crucial. It is my belief that employing symbolic representation in mathematics had a tremendous effect on the development of mathematics and furthered its application in science.

A PICTURE IS WORTH A THOUSAND WORDS

Learning Simply: The Visual Vs. the Written

In the case of certain applications, a picture is worth a thousand words. A quotidian example of this is how we rely on maps to navigate our physical world. Even in our modern world of a talking GPS providing our driving directions to our destination, your brain does not really know where you are, until you take a look at the visual map. A similar example shows up in the legal description of property. Compare the description of a plot of land shown visually/geometrically (Figure 10) and again described in text (Figure 11).

At a glance, from the visual diagram, one can deduce, two adjacent parcels of land and their general shape (trapezoidal). In contrast, the legal description in text, while being quite detailed and precise, becomes virtually impossible to grasp even in multiple readings.

These cases show that our neural network is strongly associated with visual perception proving that a picture is worth more than thousands of words.

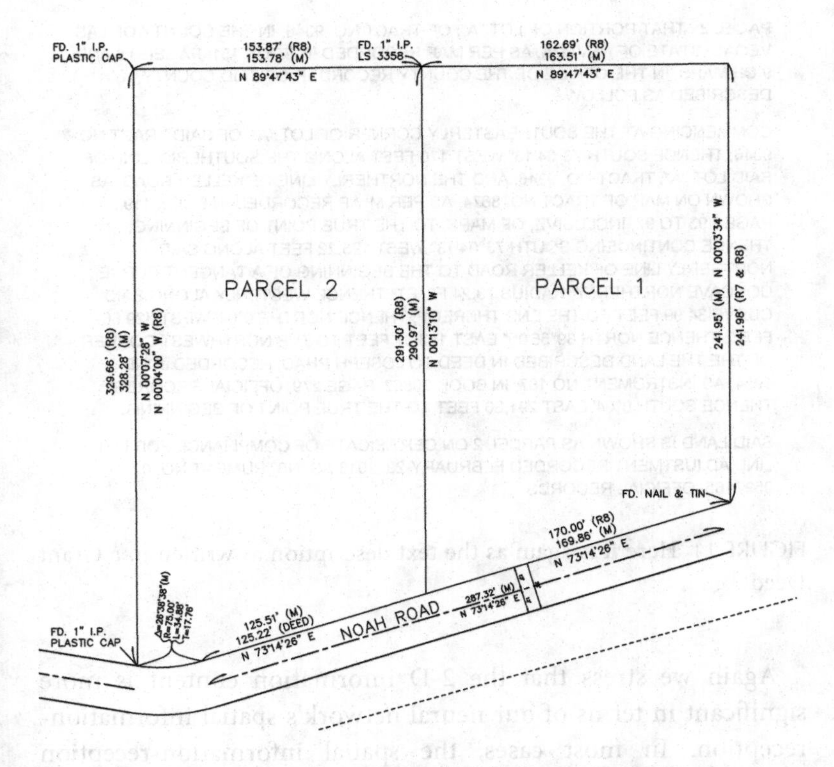

FIGURE 10 Here is property visually described in a Grant Deed

LEGAL DESCRIPTION:

PACEL 1: THAT PORTION OF LOT "A", OF TRACT NO. 9346, IN THE COUNTY OF LAS VEGAS, STATE OF NEVADA, AS PER MAP RECORDED IN BOOK 151. PAGE(S) 8 AND 9 OF MAPS, IN THE OFFICE OF THE COUNTY RECORDER OF SAID COUNTY, DESCRIBED AS FOLLOWS;

BEGINNING AT A POINT IN THE EASTERLY LINE OF SECTION 6, TOWNSHIP 1 SOUTH, RANGE 16 WEST, SAN BERNARDINO MERIDIAN, WHICH POINT IS THE NORTHEASTERLY CORNER OF TRACT 8674, AS PER MAP RECORDED IN BOOK 119, PAGES 93 TO 97, INCLUSIVE, OF MAPS IN THE OFFICE OF THE COUNTY RECORDER OF SAI D COUNTY; THENCE NORTH 0°56'05" WEST ALONG THE SAID EAST LINE OF SECTION 6,241.98 FEET' THENCE SOUTH 89°56'05" WEST 162.69 FEET; THENCE SOUTH 0°04' EAST 291.30 FEET, MORE OR LESS TO A POINT IN THE NORTHERLY LINE OF KELLER ROAD, DISTANT THEREON 170 FEET FROM THE EAST LINE OF SAID SECTION 6; THENCE NORTH 73°04'13" EAST ALONG SAID NORTHERLY LINE 170 FEET TO THE POINT OF BEGINNING.

SAID LAND IS SHOWN AS PARCEL 1 ON CERTIFICATE OF COMPLIANCE FOR LOT LINE ADJUSTMENT RECORDED FEBRUARY 29, 2013 AS INSTRUMENT NO. 02-0597165, OFFICIAL RECORDS.

PACEL 2: THAT PORTION OF LOT "A", OF TRACT NO. 9346, IN THE COUNTY OF LAS VEGAS, STATE OF NEVADA, AS PER MAP RECORDED IN BOOK 151. PAGE(S) 8 AND 9 OF MAPS, IN THE OFFICE OF THE COUNTY RECORDER OF SAID COUNTY, DESCRIBED AS FOLLOWS;

COMMENCING AT THE SOUTHEASTERLY CORNER OF LOT "A", OF SAID TRACT NO. 9346; THENCE SOUTH 73°04'13" WEST 170 FEET ALONG THE SOUTHERLY LINE OF SAID LOT "A", TRACT NO. 9346, AND THE NORTHERLY LINE OF KELLER ROAD, AS SHOWN ON MAP OF TRACT NO. 8674, AS PER M AP RECORDED IN BOOK 119, PAGES 93 TO 97, INCLUSIVE, OF MAPS, TO THE TRUE POINT OF BEGINNING; THENCE CONTINUEING SOUTH 73°04'13" WEST 125.22 FEET ALONG SAID NORTHERLY LINE OF KELLER ROAD TO THE BEGINNING OF A TANGENT CURVE CONCAVE NORTHERLY, RADIUS 73.24 FE ET; THENCE WESTERLY ALONG SAID CURVE 34.99 FEET TO THE END THEREOF; THENCE NORTHE 0°04' WEST 329.66 FEET; THENCE NORTH 89°56'05" EAST 153.87 FEET TO THE NORTHWEST CORNER OF THE THE LAND DESCRIBED IN DEED TO JOSEPH PHAD, RECORDED JUNE 9, 1954, AS INSTRUMENT NO 167, IN BOOK 20862, PAGE 279, OFFICIAL RECORDS' THENCE SOUTH 0°04' EAST 291.30 FEET TO THE TRUE POINT OF BEGINNING.

SAID LAND IS SHOWN AS PARCEL 2 ON CERTIFICATE OF COMPLIANCE FOR LOT LINE ADJUSTMENT RECORDED FEBRUARY 29, 2013 AS INSTRUMENT NO. 02-0597165, OFFICIAL RECORDS.

FIGURE 11 Here it is again as the text description as written in a Grant Deed

Again we stress that the 2-D information content is more significant in terms of our neural network's spatial information-reception. In most cases, the spatial information-reception somewhat helps our neural network remember longer and understand the learning more efficiently.

E = MC²

Learning Simply: Einstein's Intention of Simplicity

It would be difficult to remember and to understand the energy equation of Einstein if it were represented by an overly complicated equation!

Let us review Einstein's famous energy equation:

$$E = mc^2 .$$

It can be derived by using a little bit of algebra. Stating with Einstein's special theory of relativity the mass of a particle can be written as,

$$m = \frac{m_o}{\sqrt{(1 - v^2/c^2)}} = m_o(1 - v^2/c^2)^{-1/2}$$

In view of binomial expansion the above equation can be written as,

$$m = m_o\left(1 + 1/2\frac{v^2}{c^2} + \text{terms of order } \frac{v^4}{c^4}\right)$$

If we multiplied the preceding equation by the speed of light c^2 and, noting that the terms with the orders of v^4/c^2 are negligibly small, the above equation can be approximated as,

$$m \approx m_o + 1/2\, m_o v^2 \frac{1}{c^2}$$

It can also be written as,

$$(m - m_o)c^2 \approx 1/2\, m_o v^2$$

The significance of the preceding equation is that $m\text{-}m_o$ represents an increase in mass due to motion, which is the kinetic energy. And $(m\text{-}m_o)c^2$ is the extra energy gain due to motion. What Einstein postulated is that there must be energy associated with mass even at rest. And this is exactly what he discovered in his famous energy equation,

$$E \approx mc^2$$

Where E represents the total energy of the mass and

$$E_o \approx m_o c^2$$

is the energy of the mass at rest,
and $v = 0$ and $m = m_o$.

This is one of many stimulating examples demonstrating the simplified energy equation of Einstein. We can inmagine if the equation would have included all the insignificant terms, the equation would be more complicated to understand, to remember, and to apply. In fact, I believe Einstein had the intention of creating a simple and elegant expression of his theory from the very start.

> Genius might be the ability to say a profound thing in a simple way.
>
> Charles Bukowski

SYMBOLS

Learning Simply: Mathematical Symbols

Symbolic representation was developed throughout civilization to help with learning and evolve mathematics into more complex systems. One can think of mathematical symbols as part of a language: Certain symbols synthesize a function or action of one thing onto another, much like a verb describes the action of a subject on an object.

1,234,567,890 ∫ π λ Ω μ φ & Σ *f* (*x*) π + -/x ≠ ≈ (*a*,*b*) etc.

Without these simple symbolic representations, mathematics and science would be more difficult to facilitate, to develop, to learn, to understand, and to apply!

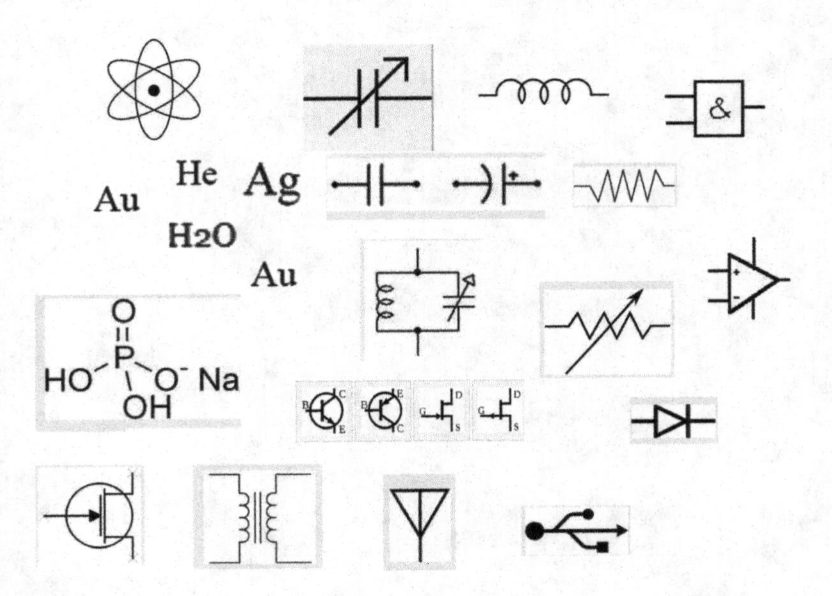

FIGURE 12 These symbolic representations simplify the learning of science and engineering

We further stress that the usage of symbolic representation in mathematics is very crucial. I believe that the simplicity of the symbolic representation in mathematics has had a tremendous effect on the development of mathematics itself and its application to the sciences.

CHAPTER 5

Memory Tricks

All this is to ask the question: must we deeply and completely understand and learn everything? Is it even possible? There certainly are times when true understanding, or even basic understanding, is just too difficult for one reason or another (due to time constraints, mass quantity of information required to remember, etc.). In such cases, I would suggest memory tricks. Unless you are a HIGHLY talented audio learner, I suggest using associative memory devices to help you along. We've seen that it just takes some learners a little longer then others to understand a concept, but eventually, with some effort, they will.

ASSOCIATIVE MEMORY LEARNING

We know that our brain stores and retrieves information and is frequently aided by the process of association, and especially visual association. As previously noted, learners can skew more towards one or the other of the visual vs. audio learning spectrum. I believe both types can take advantage of tricks that use association to aid memory.

For example, when remembering Chinese or Japanese names, we primarily use the **meaning** of the names rather than remembering the phonetic sounds, since each word of a Chinese or a

Japanese name represents a meaning and is in fact *pictorial* in nature (Figure 13).

In contrast, English names are mostly adopted from Christian names. While it is true that Christian names are associated with Biblical meaning, we generally remember these names phonetically (that is, in the modern multi-cultural world we live in now).

In a typical United States classroom of fifty students, there is easily the possibility of having more than two students named Chris. Whereas in a Japanese or Chinese classroom, each individual name carries a unique and specific meaning and visualization. Inherent to these two Asian languages is a *pictorial visualization* more easily recalled in the neural networks. Perhaps we can transfer this visualization technique in recalling Anglo names as well. Indeed, it is a commonly known technique to use some sort of visual association of a common object with a newly introduced person to more easily recall and remember this person's name in the future. To note: These pictorial-based Asian languages do indeed have other limitations, which the Roman/Anglo language excels at and this will be demonstrated later in this book. Basically, each language has its inherent pros and cons and if we can recognize this, we can benefit from the strengths of both!

We know that the human brain which stores and retrieves information is aided by the process of association.

A couple of examples:

1. Remembering Chinese or Japanese person's names are usually based on the association of meaning rather than the sounds (i.e. phonetics)
2. Memorizing telephone numbers is generally based on sequencing, symmetric properties, rhythm

FIGURE 13 Memory-based process of association

Keeping in mind that our modern day of smartphones and electronic devices do this work for us, bear with me for a moment for an illustration of how to remember long passages of numbers. If we partition the phone numbers first, and then look at the symmetric property and the sequencing of the numbers, or even the rhythm of the phone number, it is not difficult for anyone to remember more than thirty phone numbers. In fact, there is science behind the seven-digit number. Understanding that we humans have a finite limitation to our short-term memory, psychologist George Miller, in 1957, dubbed this as the "magical number seven." Basically, the short-term memory in our brains are best able to recall things up to seven digits at a time.

To best remember newly introduced names or places (whether foreign or common to one's native language), one may associate these names with an object or place that you are familiar with. These are some associative memory examples to improve our memory capacity.

For example, let me elaborate to you a story of how I remembered Professor Kazuo Nagakawa's name.

The first time we met, he gave me his name verbally. Since I neither spoke nor read Japanese, I could not remember it. Subsequently, on many occasions at various international conferences, this Japanese scientist always greeted me by my name. It was embarrassing for me to constantly ask for his name. Finally, he gave me a business card with Japanese characters which are identical to Chinese characters (called Kanji). Then suddenly I saw in those characters that his name means "a gentleman swimming in the middle of a river," as can be seen in Figure 14. After that, I easily remembered his name. Even after not seeing him for many years, I still easily remember (and visualize) his name to this very day.

We can see that, in terms of memory aid visual associative memory is one of the best techniques for the storage and retrieval of information (Figure 15).

Examples:

Japanese Name: **Kazuo Nakagawa**

Characters: **中川一夫**

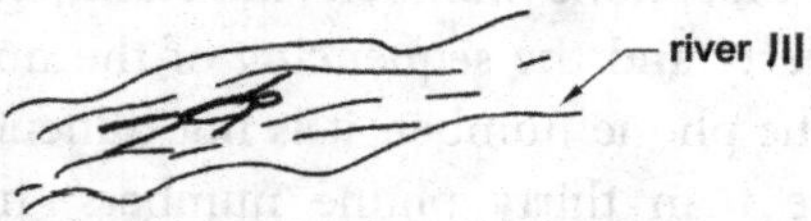

river 川

Meaning of the characters is "a gentleman is swimming in the middle of a river"

Chinese Name: **Shi-Tien Pao**

Characters: **包事天**

Meaning of this name is Mr. Pao works for Heaven.

FIGURE 14

Associative Memory is one of the best techniques for storage and retrieval of information in a neural network

Reasons:

1. Capable of retrieving information from partial or erroneous input.
2. Robust even with noise and distortion interference
3. Ability to select the closeness or similarity of information

FIGURE 15

Visual associative memory is strong due to:

- It's capability of retrieving erroneous or partial input by remembering holistically (rather than sequentially). For example: a partial image of a letter or a missed printed word can be recognized and corrected or filled.

- It is very robust under heavy noise or distortion. When detecting an object or face in a foggy atmosphere, we generally can still grasp who or what that thing is.
- It has the ability to identify the closeness or similarity of an object or information. We are able to compare and contrast quickly and fairly accurately evaluate, and ultimately identify a thing/being. For example, when looking at similar species we can quickly identify a **cat** from a **dog**.

In another potentially interesting example of memory aided by association, a few months ago I saw a segment on the CBS program, *60 Minutes*, about a young boy in the United States who can remember over a hundred digits of pi (π)! Immediately, I felt that there must be a method the young boy used to remember this sequence of numbers. I started to wonder, was he associating the sequence of digits with music, or a picture, or did he permute the sequence into arrays of numbers?

As we know, **every problem has multiple solutions**. One of the solutions we discovered is given as follows:

Let us start with π as a sequence of numbers shown below:

$$\pi = 3.14159265358979323846264338327950288419716939937$$
$$5105820974944592307816406282089986280348253421 1$$
$$706798214808651328230664709384 4\ldots$$

If I tried to actually remember this sequence of numbers as it is it would be a daunting task.

What if I permuted the sequence into a list of ten-digit arrays?

3141592653	1693993751	9862803482
5899793238	0582097494	5342117067
6264338327	4592307816	9821480865
9502884197	4062862089	

And if we further partition these arrays into a format shown below:

314 – 159 – 2653 169 – 399 – 3751 986 – 280 – 3482
589 – 979 – 3238 058 – 209 – 7494 534 – 211 – 7067
626 – 433 – 8327 459 – 230 – 7816 982 – 148 – 0865
950 – 288 – 4197 406 – 286 – 2089

it is apparent that this array of numbers is essentially equivalent to eleven telephone numbers! If we try to remember eleven telephone numbers using associative memory techniques we find the task to be entirely possible and perhaps even easy. Therefore, we have changed what seemed impossible to remember into something that is not only possible, but fairly easy for anyone to remember.

Thus we see that, by using a smart associative process, one can improve the storage capability of a neural network. And it is in fact more fun to learn how to remember things this way!

Remark

The example of remembering a hundred digits of pi (π) shows one of the inherent capabilities of a human brain: it can process two-dimensional information formats more efficiently than one-dimensional representations, a subject that we will discuss in Chapter 10.

Finally, one last example of associative memory was a trick I used during a trip to northern Europe. I had taken on the adventurous idea of driving as my means of travel, much to the consternation of my wife! Being new to the area we were not at all familiar with the various languages, let alone the complicated foreign street names. Thus, the street names seemed unusually long for us foreigners, as well as being quite impossible to read at normal driving speeds. As a means of survival on the open road, I took to boiling the process down to two simple easily digestable parts: first, I endeavored to memorize the first letter of the street name we were looking for; and second,, I prioritized my brain to remember the approximate size of the street names (small, medium, large, extra large). These simple associative processes alleviated much of the language impediments and proved that I could somewhat competently drive through most of the European continent.

In summary, by using a smart or associative learning device, we can increase the storage capacity dramatically in a neural network, making the learning process a much more efficient and fun way for remembering things. To further expound on associative learning processes besides visual associations you could enlist spatial and temporal devices in your brain. In fact, we often use the spatial and temporal when we are driving and finding our way via built landmarks ("take a right at the red gas station"), and geographical landmarks ("drive north towards the San Gabriel Mountains"). These are associative devices we all have at hand in our brains. It only takes a bit of creativity and fun to access and utilize them.

One comment I would make is that by overburdening children with extra homework we may create several negative consequences. One simple fact is that biological neural networks (i.e., our brains) need time to rest and to relax. To push children well over the curve is not a smart way of training their neural networks! Remember, as you gain something the same time you may also lose something. Allowing your children to interact with others is also a part of life education. By overburdening them with extra homework that is not required in the educational system, just to push them well over the curve is, in the long run not healthy for children mentally or physically!

> The more you try to remember the more you *cannot* remember. Then one should not try to remember *too much,* otherwise one would be *wasting* most of one's efforts.

CHAPTER 6

Talent

Few people are actually born geniuses. Genius has to be nurtured. Sometimes, however, sometimes we mistake a better prepared youngster for a genius.

For example, if Mozart was born a century earlier, or if he had been adopted into a poor farming family, his musical talent might not have been developed or discovered!

Some behavioral geneticists estimate that heritable characteristics only account for about 15 to 20 percent of the total variance between people. A person with certain inclinations has to have suitable environmental factors, such as training, practice, and timely opportunities to become a "genius."

For average youngsters who are not born geniuses, statistically they can improve their neural network processing abilities by smart learning through innovative training or education.

Our human brains are metabolic and plastic and history has shown us some of the great scientists, mathematicians, scholars and artists, were not born geniuses. It was in later years that they evolved to acquire that trait! One of them is the one who developed the famous energy equation, $E = mc^2$.

ALBERT EINSTEIN

When we read Einstein's biography, we realize that he had a child's wonderment of the world which persisted throughout his life. This continuing curiosity, plus good education and persistence effort, eventually made him one of the greatest scientists in history. Einstein studied physics and worked in the Swiss patent office where he had time to conceptualize and write the theory of relativity.

CHAPTER 7

Curiosity and Discovery

Most of the great discoveries came about from an individual being particularly curious. Einstein was a particularly curious individual and so was Leonardo da Vinci. Often you'll hear a scientist or inventor say that an accident led to a discovery. But they are being humble. The fact is that their relentless curiousity made the accident itself become an opportunity and thus a gateway to a truly new discovery. In fact, I believe "accidents" are happening before our very eyes every day. It's the truly curious individual that can take advantage of these "mistakes" and turn them into an opportunity for discovery.

For example, several thousands of years ago, we humans became aware of the phenomenon of the weight of an object. So why did it take thousands of years for the laws of gravity to be discovered by Newton? The answer is **curiosity!**

> I have no special talents. I am only passionately curious.
>
> Albert Einstein

CHAPTER 8

Unsupervised Learning

Let us move on to the issue of supervised learning (SL) and unsupervised learning (UL), as defined in Figure 16. For example, when a teacher organizes materials and then presents these materials to students, it is known as supervised learning. The danger of

<u>Supervised Learning (SL)</u>

- Teacher organizes the information and then instructs the students
- This model often requires a more passive way of learning where the student is directed from start to finish by teacher

<u>Unsupervised Learning (UL)</u>

- Student learns by themselves and independently. Bases learning on an experiential level. Often can be project-based developed by the student.
- Often a trial-and-error method where the individual learns via developing and adapting their own process to navigate problems and defining solutions (as well as defining what the problem is in the first place!). Takes Creative Adaptability to the forefront.

FIGURE 16 Supervised learning and unsupervised learning

constantly providing our students with only supervised learning is that it slips the student more easily into becoming a **passive participant vs. an active learner.** On the other hand, when students learn by themselves, whether through trial and error, or in any other capacity, the source material is more likely to become embedded into the brain with the growth of new neural networks. I believe this approach (unsupervised learning) is the best way to learn. Likewise, it is most important to understand that each individual has a different process of learning and we need to give them the latitude to explore and find their own way, in a process that suits their own individual, and unique, brains the best.

Don't assume
More knowledge equals
More intellegence

When I was in graduate school there were many courses (supervised learning) I did not take. Either due to a scheduling conflict or other reasons, I was not able to take these courses. The learning assumption at that time was that the more courses you took the better you were prepared for a job after graduation. However, in reality it was impossible to take all the courses even if I had time or would like to take them. Nonetheless, I disagreed with the assumption that the more courses you took the better prepared you would be. The fact is that one cannot be a professional student in life. As I mentioned before, our biological neural networks (i.e., our brains) are not designed to remember a lot. Instead of overwhelmingly loading our neural networks (taking too many courses at the same time), we should enjoy the benefits of our neural plasticity. Learning is a lifelong process and should not be restricted to simply taking courses. We *can* and *should* be encouraged to learn on our own.

For instance, some of my colleagues may know that I never took a single class in optics, information theory, communication theory, probability theory, stochastic processes, or abstract mathematics. Yet I have written many papers and books in optics and photonics.

I also have contributed a couple of books and papers to communication and information theory. I have written papers in applying stochastic models to photon distribution for night vision devices, as well as to photographic noise distribution, and others.

How I did I do it without actually taking these courses? The answer is simply that I learned them by myself (i.e., unsupervised learning). My experience was that it was more fun and more efficient compared with actually taking the courses. With no pressure and no exams, unsupervised learning can provide a more efficient and effective way of learning.

Why is self-learning more efficient? The answer is that our neural networks (our brains) have the capability of learning based on learned rules and past learning experiences which select the essence of what is to be learned without getting lost in the details. In contrast, when taking a course we have to remember the details and the complexity of the subject, which can be forgotten very quickly. If we did not prepare for these complexities and the details that we needed to remember to pass the exams we might not even able to pass the course! In actuality, it is more time consuming than to learn by ourselves.

Secondly, some of the most fearful aspects of a course a student must face are quizzes and exams. Even to this day, I still have nightmares about facing examinations. During those days, it was a prevalent notion (and is still prevalent today) that quizzes and exams determine what a student has actually learned.

By championing self-learning, however, we do not have those kind of fears! We can learn the materials at our own pace and at our own pleasure. We can select, learn, and *discover for ourselves*. In this way, we do not rely on the teacher to point out details and tell us what is important. We do not need to worry about grades and examinations! In fact, in using this approach, we make the necessary neural connections that commit our brains to true learning. You select what is interesting and good for you to learn. These conditions make learning more fun, more effective, and more efficient!

Of course, unsupervised learning does not preclude course work. In elementary and high school students are generally not mature yet. Most students need supervised learning. The question, though, is not just one of supervision but how can we creatively and dynamically use a more innovative supervised teaching technique which integrates the teacher and the student in ways that help the student *teach him or herself.* Regardless, unsupervised learning is still the more efficient way of learning for all ages! What I want to point out is that an examination score usually represents an indicator of a test result and is not the absolute measure of learning!

Nevertheless, without using grades and scores, in practice it is not easy to determine one's relative accomplishment in course work! So it is up to yourself to determine what you have actually learned. At the same time, to maintain a good score and actually understand what you have learned is even better, of course. I understand the importance of getting a good grade as it can open doors to allow one to go to the next step (i.e. get into a university). However, if you can even learn and grasp one concept or thing, no matter how small or imperfectly, I would say that it would stay with you forever. In other words, what you actually learn is what you have and nobody can take that away from you! It may not be a reflection of your score, but it will eventually add up as time goes on. In fact, I have discovered other higher-level educators, professors, and PhD candidates who are able to produce volumes of calculations but are still not able to express the basic concepts behind their equations. This is NOT uncommon. Never be ashamed of what you do NOT know. In fact, embrace it!

Which leads me to the next point:

> Never be ashamed of what you do NOT know.
> In fact, embrace it!

CHAPTER 9

Mistakes

I've always told my students to not be afraid of making mistakes as you will only learn from them. The one who is afraid of making mistakes is the one who will not learn!

Shortly after I received my doctoral degree, I wanted to learn optics and phonics by myself. I bought a basic textbook in optics written by Professor Francis W. Sears at the Massachusetts Institute of Technology and quickly started learning optics by myself. At the same time, I was a consultant at a defense research institute in Ann Arbor (a college town where the University of Michigan is located) so that I could get involved in research activity in the areas of optics and photonics. Without any prior preparation, I called up the institute and asked to talk to the director directly. Although we had never met before, the director hired me as one of his consultants. I remember that he was a bit surprised to see an unknown youngster, without any experience in the area of electro-optics, seeking a consulting position! Nevertheless, he agreed to meet me in Ann Arbor (where I lived) not far from their research complex. To make a long story short, after the meeting, although I did not have an optical background, his group of scientists were impressed with my stochastic modeling. And that was the major reason he hired me on the spot. As I had noted earlier, I had never taken a course in stochastic processes, and had learned it on my own.

The consulting problem was, as I recalled, evaluating the photonic distribution within a micro-channel tube as it applied to a night vision device. Since I was not a U.S. citizen at the time, I was not allowed to see the device, due to the classified nature of the project. As I recall, what I did was draw a small tube to simulate a micro-channel phonic multiplier on a piece of white paper, with a few physical parameters added. The scientists agreed that fitting the parameters into the stochastic model was not a major problem, although the calculation was rather complicated. Basically, the problem was similar to a birth-and-death random process, which I had learned on my own.

I had to meet this group of scientists on a biweekly basis and discuss the progress of my investigation with a short report. At the end of my consulting contract (about a two-month period), I gave a formal presentation to this group of scientists which included the institute director. After the question and answer portion of my presentation however, they were very skeptical about my results (although I was very confident of my evaluations!). One of the major issues was when a single phone is launched into the micro-channel multiplier, based on my calculated result, it became a decaying (or inversed exponential) type distribution. Almost all of the institute scientists disagreed with my evaluated results because of this fundamental issue! I thought it was a very nice piece of work that I had done, but there was no way to convince them at that time, since I was not permitted to view the actual experiments, due to my lack of security clearance. Even though the institute would allow me to set up the experiment, I would not have the support to set the all tests by myself! Although I was quite certain of the outcome of my results, it felt awful to let them down as I did not come up with the kind of results that the scientists and researchers had anticipated! These were highly trained scientists who had developed sophisticated night vision devices and hardware. I felt that there was absolutely no way that I would be able to convince them that the analytical distribution I had obtained was appropriate and correct. Indeed, I started to doubt the conclusion myself.

Nevertheless, several months passed and I got a surprise call from the institute director. He kindly expressed to me, "Dr. Yu, we have experimentally tested your calculated distribution and it was indeed the kind of distribution that you had predicted!"

This story shows the significance of unsupervised learning. In this rapidly changing world of science and technology, the learner has to constantly update him/her self. One of the best methods is by self learning (unsupervised learning). It is more efficient, effective, and more fun!

Another aspect of this learning experience is that one should not be afraid of making mistakes. The person who is afraid of making mistakes is the one who won't learn. For example, when I had just newly received my doctorate degree I did not have an optics and photonic background, yet I had the courage to accept the challenge! I was not afraid of making mistakes. My rationale for this is that IF I make any mistakes, not only would it be the BEST way of learning, but I may actually make a significant discovery (whether it be personal or scientific). Either way, in this particular case I really had nothing to lose! There is always the risk of being "wrong". But is it better to always be "right" with the obvious solution? Or is it worth it to risk exploring a different path, perhaps one that might be "off the beaten path"? Who knows, this less travelled path might be the one that brings about a revolutionary discovery!

> The one who is afraid of making mistakes is the one who will not learn.

> Einstein "failed" many times but discovered the theory of relativity!

> Many a scientist can quote an equation but I discovered quite a few who still do not know what it means!

CHAPTER 10

Ask Questions

Don't be afraid to ask questions. When I taught coherence theory most of my students were engineers and I discovered that some of them did not actually understand the essence of *correlation*, although some of them used the correlation function. Added to that, some of the engineers did not actually know the meaning of, and basic distinction between, coherence and correlation. This was because coherence theory was developed by optical physicists (a whole different field from engineering). In my effort to help students understand (i.e., *cognitive learning*) the coherence theory of light, I usually started with a simple example as seen below. If you are able to grasp this simple illustration you will have a better fundamental mastery of the subject than many of my PhD students!

SCENARIO A (NO CORRELATION)

As we watch students rushing to their classes on a university campus, we see two students passing by each other without showing any sign of connection or recognition. In this case we can assume that these two students are not acquainted with each other.

SCENARIO B (SOMEWHAT CORRELATED)

On the other hand, if we see two students stop momentarily to speak to each other, then we can assume that these two students are indeed acquainted with each other.

SCENARIO C (HIGH CORRELATION)

If we see two students stopping to speak to each other, shaking hands while smiling and laughing, then we are sure they must be friends!

From these cases, it is apparent that the two students in Scenario A were **un**acquainted (no correlation). In Scenario B we can conclude that the two students must know each other (some correlation). For Scenario C we are sure that the two students not only know each other, they are also friends (high correlation).

In view of the preceding illustrations, we see that the degree of coherence is actually the same concept as the correlation principle. In most cases, coherence theory usually is expressed in terms of degrees (i.e., percentages). Whereas correlation function does not usually do that since the concept has a broader application in the physical and social sciences. Nevertheless, the two theories are similar and my illustration helped the students remember coherence theory well.

> A smart person is not necessarily intelligent. And a person who is intelligent is not necessarily smart. A person who looks smart may not actually be intelligent!

> He who asks a question remains a fool for five minutes. He who does not ask remains a fool forever.
>
> Chinese Proverb

CHAPTER 11

Approximation

Something I often say to my students is that an ounce of approximation is worth more than tons of calculation. Because in truth, all scientific equations are approximations. In fact Edwin Hubble (of the Hubble Space Telescope) has been known to say, "The scientist explains the world by successive approximations" (Figure 17).

Since science is a law of approximation, we know that many of the important equations in science are indeed approximated. For example, Einstein's energy equation $E = mc^2$, the Maxwell's equations, all of the Newtonian equations, and many others are, strictly speaking, **approximations**. (And they are very good approximations, I might add!)

> The scientist explains the world by successive approximations.
>
> Edwin Hubble

> An ounce of approximation is worth more than tons of calculation.

That being said, I would like to emphasize a very important difference between science and mathematics. Science is a law of

FIGURE 17 Edwin Hubble and the 48-inch (1.2-m) telescope later dedicated as the Samuel Oschin Telescope. (Image, Courtesy of the Archives, California Institute of Technology)

approximation while mathematics is an axiom of absolute certainty. Using **exact math** to evaluate **inexact science** does not guarantee that a solution exists within our temporal universe!

SCIENCE VS. MATHEMATICS

- **Mathematics** is an axiom of rigorous **certainty** but with some paradoxes.
- **Science** is a law of **approximation** and it requires constant revision.

CHAPTER 12

Every Problem Has Multiple Solutions

One point I would like to make is that "every problem has multiple solutions; it is the number of constraints that limit the number of solutions." For example, I was struck one day upon seeing a string of notated beads (Figure 18) that the invention of the abacus (Figure 19) was an improvement of this primitive calculating device. If we examine the abacus, one may ask why the Chinese did not continue to develop a better calculator (i.e., better solution). They stuck with the abacus for a couple of thousand years.

One of the apparent answers to this question is that the abacus was a very efficient calculator at that time and there was, therefore, no need to improve it. The Chinese were satisfied with it! In some ways, the abacus proved to be a very useful tool. In other ways, its success may have prevented the Chinese from developing an even better calculating machine.

> Ironically if a solution is found, sometimes it also prevents us from looking for a better solution!

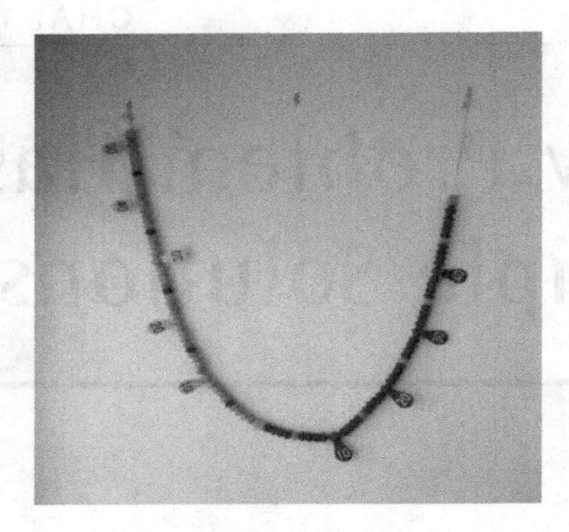

FIGURE 18

Good calculator: But can it be improved?

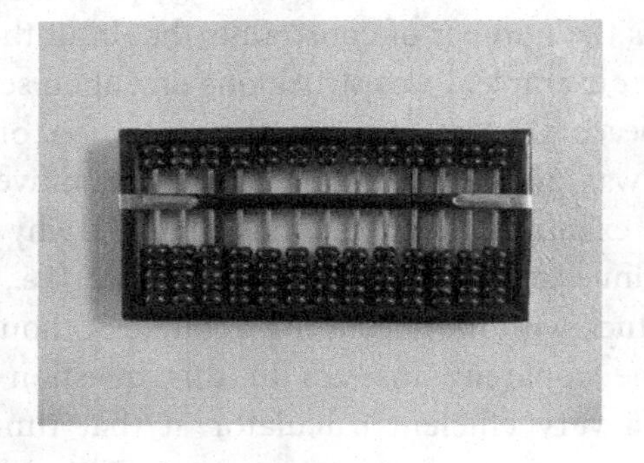

FIGURE 19

Better calculator: So good it lasted for thousands of years! (Until the more recent development of electronic calculators which eventually replaced it!)

CHAPTER 13

The Illusion of Mastery

Fast vs. Slow Learning

Fast Learning:	Slow Learning:
– Short Term Memory	– Long Term Memory
– Noisy	– Clear
– Unorganized	– Organized

PREPARATION

How much is too much? I've always told my graduate students to take the minimum course requirements and optimize their learning. They can simply audit classes or learn by themselves (i.e., unsupervised learning).

Ironically, one of the most inefficient ways of learning, from my personal experience, is actually taking a course. As we know, in order to earn a good grade, or even just to pass a course, one needs to memorize or remember a lot of details. This type of learning is what I call **fast learning.** It helps when cramming for an exam and is used quite often to obtain an A grade.

If a student wants to get an A grade, he/she has to pay close attention to tedious, complex, and redundant details. Indeed, it consumes a lot of time and effort to get that A grade.

Usually, straight-A students are afraid to learn outside the box. They are often afraid to disagree with the professor for fear of not making an A grade and may also be very afraid of making mistakes.

Since straight-A students spend most of their time chasing the ghosts of A grades, they rarely have any time left over to think about how they learned something and why!

Usually, their primary goal is to study hard and get good grades and they are not interested in learning outside the box. or by themselves, since there is no grade to chase!

Because of the unique grade point average they've acquired, straight-A students tend to be more self-centered and afraid of taking chances beyond what they were required to learn!

As we know, it takes great effort and time to prepare for, and remember, all the complex details necessary to get good grades. But since we are, in effect, overloading the natural abilities of our neural networks by learning this way, eventually we lose, not only most of the details of what we've learned, but the overriding concepts as well! It is much better to start with an understanding of the basic concepts and fundamentals, rather than the complex details, which can be forgotten easily.

Meanwhile, **slow learning** may take a more circuitous and unconventional route. This usually depends on the individual finding his/her own process of learning. It may be necessary for an Auditory-Sequential Learner to make up her own narrative to help explain something in her own authentic voice. Or it may mean a Visual-Spatial Learner might draw pictures or diagrams to explain the problem. Either way, it depends on the individual synthesizing and processing a problem in their own way and thereby reinforcing the information into their long term memory.

The Faster you learn, the Faster you forget.

Let me share a personal story in terms of overly complex routes to learning as compared with a more common sense and fun way to learn. I had a very good friend in graduate school. He was a straight-A student and even garnered a few A pluses! In our time, it was more difficult to earn an A grade in the American universities than it is today. The professors usually gave only a handful of A grades in a class (roughly speaking about 5 to 10 percent in a class). And yet, to this day, I have not seen anyone who has a higher grade point average than my friend.

This is how he earned his excellent grade point: What he did was he attended every class and meticulously read the text practically from cover to cover. He paid attention to every detail and solved every single problem in the text. He spent a lot of time solving problems. He was incredibly thorough.

My friend always prepared well ahead. He would spend hours in the laboratory, even going to the effort of preparing a few days ahead of the lab class. He would repeat over and over the same lab work, just to be sure that he would be awarded the highest grade in class. I believe his goal was always centered on his grades.

The irony was that when we were preparing for our candidacy examination, I found out that he had forgotten most of the details of his studies and some of the most basic and fundamentals concepts. Although he passed the PhD candidacy examination his scores were not as high as he had expected. This upset him tremendously because some of the lesser-grade students performed better than he did. In retrospect, I can see that his example was very typical example of rote learning (i.e., detail-oriented learning).The fact was, my friend, who was a very intelligent and hardworking student, had committed a serious error: He had became so involved in obtaining a perfect performance record that **he couldn't see the forest for the trees.**

This is an example from my real life experience that shows how our brains (neural networks) are not overly equipped to be computing machines. Our brain is good for learning the simple, the basic, and the fundamental.

> The greatest enemy of knowledge is not ignorance, it is the illusion of knowledge.
>
> Stephen Hawking

CHAPTER 14

Late Bloomers

By the time I was in grade school I already had a well-established reputation as a bad student. I was from the lower middle class and no teacher had ever shown any faith in my intellectual abilities. I had some natural ability as an athlete so I threw my energies into soccer and basketball and was seen as something of a dumb jock. Nevertheless, on a whim in tenth-grade algebra class, I decided to try my hand at solving a couple of the assigned homework problems. I did my best but wasn't sure my solutions were correct. The teacher, upon seeing that two of my solutions were indeed correct, immediately made me feel that I had copied from someone! She was patently accusing me of cheating.

The funny thing is that even though I was disappointed with her response, I wasn't surprised. If she had been more encouraging might I have changed sooner?

> The problem with the world is that the intelligent people are full of doubt, while the stupid people are full of confidence.
>
> Charles Bukowski

CHAPTER 15

Leadership in Education

Academic institutions are not like a military school such as West Point, Naval, and the Air Force academies.

Military school primarily trains students to be the leaders of the armed forces.

The primary goal, however, of an academic institution is to educate students to become good educators, scholars, scientists, engineers, artists, philosophers, and others. The success of an academic institution depends on innovative educational techniques that will enable professors in the institution to produce the brightest students after years of association with the university. In fact, the basic objective of an academic institution is not to produce students to become presidents, chief executive officers, deans, or department heads of universities. It is a wrong to advocate leadership as the primary purpose of an educational institution.

Look at the current universities across the country. With a few exceptions, the so-called leaders in education, such as the presidents, the deans, the department heads, and numerous vice presidents are actually CEOs of institutions. They function as big corporate executives! How many of them are really educators or scholars? Maybe only a handful? If I am not mistaken, it seems that the primary function of a CEO is fundraising. They build up football and basketball teams to bring in revenue. University presidents receive as high a pay and bonuses as CEOs in big companies! My question is: Are those academic-CEOs the kind of leaders in education that we want in academic institutions?

No wonder the cost of education for students is getting higher and higher! The so-called leadership in education is not actually there.

If their is leadership in an academic institution it is in the faculty, not in the CEOs of the universities!

TESTING

The more heavily our mass educational system leans on standardized testing, the less learning we are likely to see among our students. For even though test scores theoretically measure aptitude, learning, and some would even go so far as to say intelligence, if overused, testing itself can inhibit creativity, imagination, and critical thinking—the three hallmarks of human intelligence. Obviously, if you are inhibiting what you are supposed to be testing, there may be a problem. One possible reason is that testing puts pressure on a student to memorize the "right" answers, which can, in turn, prevent him or her from considering possible alternatives—what would likely be deemed as "wrong" answers by the all-too-human, sometimes machine-like, test grader.[1] In other words, pressure to be "right," as deemed by the testing authority can, and usually does, steer people away from exploring alternative questions, themes, perspectives, and subjects that could potentially lead to new discoveries, and possibly better solutions, ideas, theories, and paradigms.

Equally important, frequent testing can habituate students to rote memorization and therefore encourage obedient acceptance of "the facts." Contrast this to the thoughtful considering, questioning, wondering, searching, and ruminating that are part and parcel of deeper learning and you get the feeling that Pavlov's dogs aren't the only ones responding predictably to the master's bell. Perhaps the quandary of testing can be best summed up by Harvard psychologist Ellen Langer: "The more rigidly we learn the original information, the harder it may be to open up those closed packages to accommodate the new information."[2]

EXACTLY WHAT DO STANDARDIZED TESTS TEST?

The advent of standardized testing has resulted in a growing number of people and institutions equating test results to a literal and exact measure of complex and highly subjective abilities, aptitudes, and phenomena, even though they are, in truth, figurative measures, estimates and, at base, gut-level appraisals. Not only has our concept of general abilities and aptitudes, such as those involving thinking, learning, academic aptitude, and overall intelligence had to weather a subsequent debasement, but we appear to be experiencing a narrowing of our actual abilities and aptitudes themselves.[3] And this means both our understandings of the word *intelligence* and perhaps general intelligence itself are becoming increasingly confined to the point of caricature, as evidenced by the simple fact that, for most people, the term "IQ" actually *means* intelligence.[4]

While we may be able to literally measure very *specific* and *narrowly defined* actions, abilities, aptitudes, and phenomena, it is impossible to do so with *general* actions. In fact, it is impossible to take direct measurements or otherwise definitively quantify all but the simplest, most basic actions, abilities, and attributes and narrowly defined phenomena (i.e. those involving the fewest number of objectively measurable variables.)

When it comes to complex issues, we are therefore left with quantifying the few variables that are literally measurable and comparing the resultant values to predetermined standards and, if no such standards exist, comparing the values to personal experience before making an appraisal. How one goes about making the appraisal depends on a combination of conventions (institutional, cultural, and otherwise), unconscious personal bias and, in some cases, the kind of debate or discussion one might see in say, the Oval Office, corporate boardroom, medical, or science board, or even private in meetings with friends. Thus, while analyses of complex issues might at first appear completely "objective," "statistically-proven," "computer-generated," "fact-based," and otherwise "rational," they ultimately rely on human and, therefore, subjective, gut-based judgment.

The dichotomy between the measurable and the immeasurable is readily seen when comparing the phenomenon of air temperature to that of weather. While we can directly measure the air temperature at a specific location at any given moment, it is impossible to do the same with the weather, which is a far more complex phenomenon than air temperature and consequently involves infinitely more variables, many of which are notably unknown. To get a sense of the weather from a scientific perspective involves choosing among the known directly measurable and observable variables we want to employ, such as wind speed, wind direction, precipitation upwind from our location, barometric pressure, air temperature, humidity, and cloud cover, and then measuring and observing those variables, comparing the results to recorded standards if they exist, and then making an appraisal (using words like "mild," "stormy," "pleasant," "unseasonable," or "blizzard-like"). Even if we assign a number to our appraisal, it will represent a *figurative* measurement rather than a literal one.

Have the courage to follow your heart and intuition.

Steve Jobs

NOTES

1 Note that in the U.S., testing is a multibillion-dollar industry with standardization making robotic grading easy and profitable. I'm using the term *robotic* both literally and figuratively because companies employ literal machines to grade the tests while, at the same time, teachers' duties increasingly include tabulating ticks on a piece of paper.

2 Ellen Langer, *The Power of Mindful Learning*, Da Capo Press, 1997), p.22.

3 It is important to note that most concepts are themselves cultural constructs and therefore not entirely identical to those from other cultures and subcultures. Words consequently rarely take on exactly the same meaning when used among different groups. Popular terms like equality, freedom, justice, and democracy, for example, take different meanings depending on where you live and even what era you happen to live in. Even here in the U.S., legislation, executive fiat and court hearings continue to change the way particular municipalities, states, and the country at large defines these terms. This is why, for example, corporate sponsorship of elections has for some (mainly the wealthiest one percent of the population) become an accepted and even necessary part of their definition of the word *democracy*.

4 In fact, the abbreviation does not mean or even stand for intelligence, but instead "Intelligence Quotient," a metric designed over a hundred years ago by French psychologist, Alfred Binet, for the specific and limited purpose of identifying and helping students with special needs. The subjectivity inherent to IQ tests becomes obvious when noticing that IQ has been steadily rising over the last century—meaning that either we are geniuses or our ancestors were idiotic. For a cogent analysis on IQ testing and its pitfalls, please see Stephen Jay Gould, "Curveball," *The New Yorker*, November 28, 1994, and Stephen Jay Gould, *Mismeasure of Man* (W.W. Norton).

It should be noted that the rapid rise of standardized testing is not merely a cultural phenomenon but a political economic one. According to the *Washington Post*, "The four corporations that dominate the U.S. standardized testing market spend millions of dollars lobbying state and federal officials, as well as sometimes hiring them to persuade them to favor policies that include mandated student assessments, helping to fuel a nearly $2 billion annual testing business, a new analysis shows." — Valerie Strauss, Report: "Big education firms spend millions lobbying for pro-testing policies' — *The Washington Post*, March 30, 2015. If these corporations are raking in profits, the test prep industry could be doing even better: "As of 2015, parents spent $13.1 billion on test-prep that included preparation, tutoring, and yes, even counseling." — Rainsford Alexandra, "The Business of Standardized Testing," *The Huffington Post*, March 27, 2016.

CHAPTER 16

A Short Biography

A Little Bit about Myself

As you might have guessed, those who knew me during my formative years did not expect me to go to college, much less spend the bulk of my life in academics. Despite an outward show of insouciance and general disregard for authority, I myself wondered if my high school teachers were right in insinuating, as they often did, that I would never go anywhere in life. Their frequent looks of disdain, along with those cast down from administrators and sometimes even classmates, didn't help to build my confidence, though they did fuel my contempt for the educational system and the rigid rules its minions attempted to enforce. While in those days I sometimes blamed myself for my own "bad" behavior, I now realize how rebelling, and otherwise acting out as many children do, are not necessarily indicative of any inherent weakness in the individual. On the contrary they can, in some cases, be a normal, and even healthy response, to an oppressive system.

Despite my reputation as a flunky and troublemaker in high school, I did manage to graduate, and did it on time—albeit as an illustrious, third-from-the-bottom of my class. Of course, I was well aware that my diploma served more as a charitable gift than any sort of acknowledgment of academic competence. (Seeing how that piece of paper was the most expedient and painless way for the school to rid themselves of an irreverent troublemaker, I realize in retrospect, that I wasn't the only one benefitting from this apparent gift.) Given the circumstances, I not only felt undeserving and even ashamed on the day of my graduation, but carried a gnawing sense of anxiety and doom. How was I ever going to take care of my father's growing debt, create opportunities for eight younger siblings, and otherwise get out of the poverty treadmill if I didn't even have a decent high school education? Having faked my way through the classroom for several years, I simply felt unprepared for surviving in the world, much less helping my family climb out of its financial predicament. In my mind, I would never amount to anything and the consequences would be far reaching.

My teachers seemed to agree. Even on the day of our high school graduation, one went so far as to discourage me from even enrolling in college. Unlike the more studious classmates in whom she took a far greater interest, heartily congratulating them on their good grades and successful entrance to prestigious universities, I would be better off, she told me, continuing to work odd (and therefore, low-paying) jobs. Given this dim view of both my character and ability, it shouldn't surprise anyone that many years after this unhappiest of graduation days, classmates and teachers from this era would unfailingly express doubt, if not suspicion, that I had even attended a university. Doubt would often turn to utter disbelief if and when they discovered that I ended up, not only graduating first in my class at my university, but receiving a partial scholarship to attend graduate school in the United States, and eventually becoming a scientist, university professor, and director of a major laboratory.

During the years that I served as a university professor and research scientist, I continued to face disbelief, although this time it was of a more benign and respectful type. People who related to my formative years, those who first knew me as a flunky, thought I was too dumb, immature, and undisciplined to make something of myself. Colleagues and students who were part of my life as a tenured professor—i.e., those who only knew me as a scientist—thought I was too creative, hard working, and accomplished to have ever been a poor student. In short, those who were part of my new setting had trouble understanding how a troublemaker, born to a poor and mostly uneducated family in a feudal society, could go from flunky to scholar, all while receiving little in the way of parental, financial, or remedial support. With few exceptions, both colleagues and students who were part of my new surroundings were from well-educated and often prosperous families. They had consequently received both parental pressure to become academically successful as well as parental support to make that success highly probable. Virtually all had grown up, in other words, in families that not only supported intellectual development, but possessed the material wealth to make it happen.

In contrast, I was born in a farming village in Southern China to an illiterate 15-year-old mother and a semiliterate 16-year-old father who had only completed the first few years of grade school. When I emigrated, at the age of five, with my father to the Philippines I would be left exclusively in his care (some might say neglect) for almost a decade. Since my father was usually physically absent from home during the day, presumably off somewhere conducting business, but, as I was later to discover, often gambling and emotionally absent, except when venting his explosive temper during his brief visits in the evening, childhood turned into a lonely and sometimes precarious period in my life. Lacking parental attention, guidance and, in some cases, protection as a child, I may have indulged in more "childish" mistakes than the average kid. Unfortunately, childish mistakes made shortly after the end of the war, and in one of the larger ghettos of Manila could have injurious, if not lethal consequences. It didn't help that I was constantly dealing with neighborhood bullies and couldn't resist their challenges, even though this invariably ended in my being trounced. The danger inherent to juvenile street fights, however, paled in comparison to my growing interest in explosives. Shortly after the war ended, I discovered a new toy in the form of an unexploded shell leftover from US and Japanese occupying forces. Unlike fighting, however, my interest in them was somewhat short lived, ending on the day I saw one explode in my neighbor's face, rendering him permanently blind.

Despite all of the negatives, lack of parental supervision did have its perks. namely the freedom it afforded me to figure things out on my own and consequently, discover the "pleasure of finding things out" (to steal an apt phrase from Richard Feynman). It was a gift I would carry with me for the rest of my life. For learning at one's own pace and directed towards one's own interests is not only pleasurable but, as I was to realize later, the most effective way of expanding one's mind.[1]

When my mother was finally able to join us during my teen years, I was already far too rebellious (or "wild" as my teachers put it) to be controlled and otherwise turned into an obedient student. Being the oldest of what eventually turned out to be nine children, I had the responsibility of helping to provide for the family, which at times meant keeping my father's creditors at bay and finding ways to slow the rise of his, or rather *our*, perpetually growing debt.

I was not failing in all aspects of my life, however. For as much as I displayed little academic aptitude, I excelled in sports. This may be one reason why, early on, I became more interested in playing soccer and basketball than doing homework, or even showing up for school at all. My interest grew out of, not only the physicality, but the joy of searching for new patterns of movement and strategies for overcoming opponents' defensive maneuvers. Later I would realize that imagining and experimenting with new ways to move past defenders and score goals or baskets (more specifically, new ways to move my body and control the ball and, all the while, putting both into favorable positions relative to that of my opponents), was not dissimilar to imagining and experimenting with ways of solving any other problem, whether it be in mathematics, science or such mundane tasks as fixing the kitchen sink.

Becoming a star on the local soccer team helped to make up for the insecurities I harbored for being labeled a flunky and troublemaker. Teachers and classmates didn't have to respect my academic performance, which was admittedly poor, but nobody could say I was a bad athlete. Through sports, I was finally able to garner respect from peers— albeit primarily from tough guys and fellow flunkies—and, consequently, some modicum of self-respect.

As with many athletes I knew at the time, schoolwork was always secondary, if it was any consideration at all. Of course, for me, this greatly reduced the odds of getting out of poverty, the minimum prerequisite for which I needed a decent education. As a result, I had a growing sense that I was reaching the end of a road—one from which I had to quickly veer off if I was to avoid living the rest of my life as a menial laborer, bum, hustler, or petty criminal. Moreover, this kind of ending would not bode well for my family, all the members of which were obviously stuck in the same treadmill of poverty as me. Thus, the urgency of getting a decent education and making something of myself grew with each passing year and each failing grade.

In the end, and to the surprise of everyone around me, I was able to turn things around once I enrolled in college. Of course, the change seemed miraculous to outsiders, but understanding the learning path I had followed from an early age, it is not entirely outlandish. In fact, the first and seemingly unorthodox steps I had taken in grade school, while providing little in the way of acceptance or success in the classroom, turned out to be formidable training for a life in academics, science, and research.

Looking back, I was one of the very lucky few who were able to overcome a relative lack of privilege and what my well-bred teachers and peers might have called delinquency, to not only acquire a good education, but succeed in the academic world. Consequently, I have received countless queries from students over the years about my "secret learning methods." In truth, none of my methods are a secret, though they probably appear as such because they stray so far from convention. For, as you shall see in the following pages, there is much about convention and orthodoxy that holds people back from learning. Because learning has become something of a lost art, this book is an attempt to reveal what I consider universal methods to learning—what some people call "the lost art of learning." In doing so, I will first uncover what it means to truly learn (as opposed to mimicking, pretending, parroting, or otherwise acting in robotic ways) and, as you shall see, what it consequently means to be human. For in order to learn better, or learn at all, we must first know what learning is. And to know what learning is, it will help to understand what it means to be human.

–by Edward Yu (as told by Francis Yu)

NOTE

1 The dearth of such self-directed exploration may explain why, in this hurried modern age, there is so little learning happening in our schools and universities.

Appendix

For much of my career I conducted research with a team of graduate students in my grant-supported, electro optics laboratory that I created from 1980–2005. The following sections provide a glimpse of some of the actual experiments that we conducted and published in scientific journals. These experiments utilized an optical neural network that my team and I developed in the search for an artificial neural network.

A. ARTIFICIAL NEURON OPERATION

Let us now turn to the artificial neuron operation. Again, I define the biological neural network as our human brains. Likewise, the artificial neural network is our human attempt to simulate a close facsimile to the biological neural network (BNN) and is akin to a computer. In general, a neural network of N neurons has NxN interconnections. The transfer function of a neuron operation can be described by a nonlinear operation (e.g., a step function), making the output of a neuron operation in a 0 or 1 binary state, or a sigmoid function that gives rise to an analog value.

Let us now present an artificial neuron operation as shown in Figure A1.

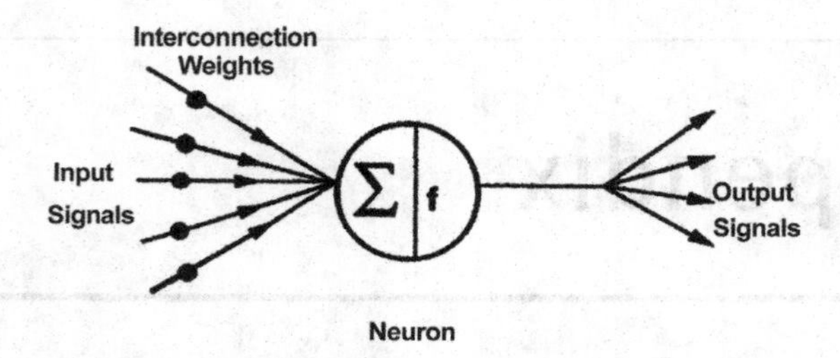

FIGURE A1 Artificial Neuron Operation

For the N-neuron neural network the output iterative equation can be expressed as,

$$V = f\left[\sum_{i=1}^{N} T_i X_i\right]$$

Where T represents a two-dimension memory (also called iterative) matrix, N represents a linear array of neurons, and f is a nonlinear operator.

It is therefore apparent that in an NXN array neuron network the iterative equation can be represented by a two-dimensional format as shown in Figure A2.

Thus we see that for a one-layer, NXN-neurons network, the memory matrix can be represented by a two-dimensional format which is very suitable for panel optical space interconnection.

Notice that the two-dimensional memory matrix T can be treated as a two-dimensional artificial brain to facilitate subsequent demonstrations. The two-dimensional matrix formulation allows us to develop an optical neural network as we will show in a moment

Further note that for a two-layer neural network the number of interconnections is equal to the square of the number of neurons.

A 2-D Neural Network Representation

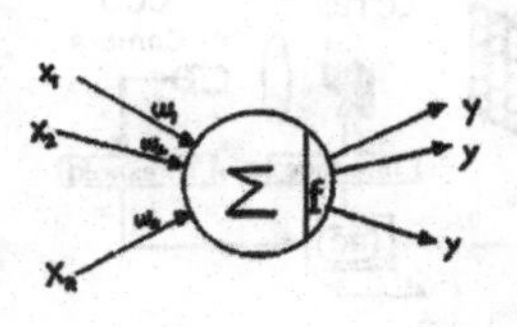

$$V_{lk} = \sum_{i=1}^{N}\sum_{j=1}^{N} T_{lkij} V_{ij}(n),$$

$$V_{lk}(n+1) = f(V_{lk}),$$

where T_{lkij} is a 4-d matrix, which can be partitioned into NxN sub-matrices, and each sub-matrix is NxN size.

$$V = f\left[\sum_{i=1}^{N} T_i X_i\right]$$

Iterative equation

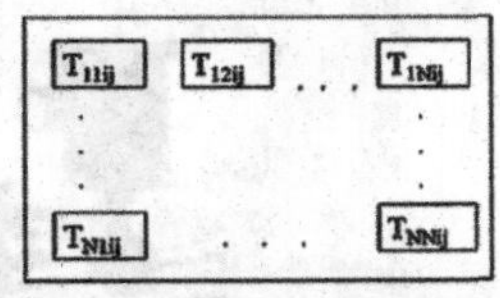

Thus a 4-D T can be expanded into an array of 2-D sub-matrices.

FIGURE A2 Neuron Operation

For example, a fully interconnected 10^4 neurons network requires about 10^8 interconnections. This magnitude is difficult to simulate and beyond current, state of the art Very Large Scale Integration (VLSI) technology.

Optics offers the advantages of parallel processing and massive interconnection capabilities for the design of a large-scale optical neural network. In Figure A3 we show an optical neural network architecture, developed at Penn State. The upper part of the slide shows the schematic diagram, while the lower part shows the experimental setup.

We note that it is not our objective to describe the design and performance of this optical neural network. (For those readers who are interested we refer them to F. T. S. Yu and S. Jutamulia, *Optical Signal Processing, Computing, and Neural networks*, Wiley-Interscience, New York, 1992.) It is, however, our objective to show some of the results that were obtained by this optical neural network in order to facilitate our demonstrations.

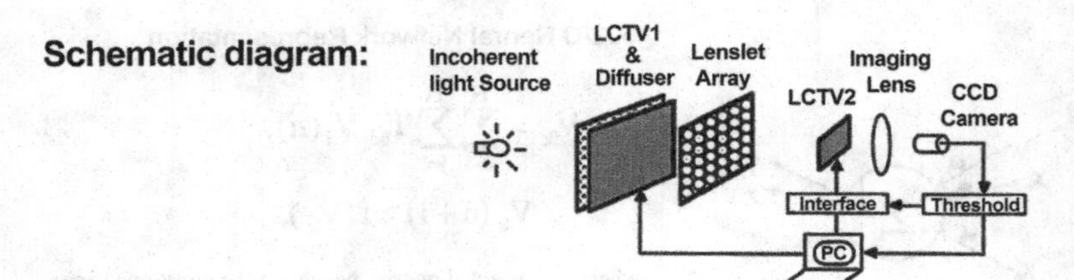

Experimental Setup:

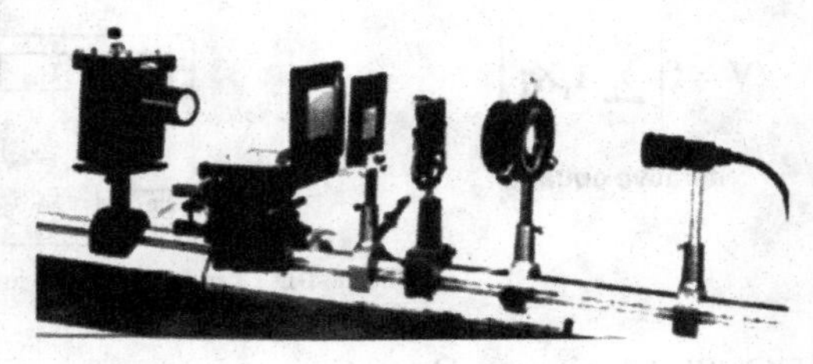

FIGURE A3 Penn State Optical Neural Network

A question, however, remains to be asked. Why use optics for the development of a neural network (Figure A4)?

Our response is that optics offers the advantage of:

- high-speed processing capability
- high-density space connectivity
- capability of performing parallel processing, as well other things.

An optical system, however, is still not very flexible to manipulate. For this reason, the optical architecture we developed was actually a hybrid system (i.e., with computer invention). In this respect, we were able to take advantage of both the analog and the digital to a combined, better effect.

We emphasize that, although an artificial neural network operation is still far from the reality of a biological neural network, it is one step closer to its realization. In fact, the two-dimensional representation of the memory (or interconnection)

Optics has the advantages of
• massive interconnectivity
• high density
• parallel proccessing capability
• and others
However, optics is still not very flexible to operate.
Therefore a hybrid-optical NN is more suitable.

FIGURE A4 Why Optical Neural Network?

matrix **is very suitable for the presentation of the neural network's adaptability, storage capability, as well for fast-learner and slow-learner neural nets.** We shall discuss them in more detail as we encounter them in subsequent sections.

B. ASSOCIATIVE MEMORY NEURAL NETWORK

Let us now move to the associative learning neural network. We shall now utilize the optical neural network to facilitate our demonstration as shown in Figure A5.

In this figure we used A, B, C, D English letters as our training set. In other words, if we presented this set of letters repeatedly to the optical neural network, which has been pre-programmed as a neural net, then the optical neural network remembered these four letters. If we presented a partial letter "A," as shown at the input end—that is on the lefthand side of the optical neural network—then we obtained a completely retrieved letter "A" at the output end (i.e., on the righthand side) of the optical neural network. Because of its use of associative learning, the optical neural network was able to easily grasp an error or omission and correct it. Thus we see that the optical neural network can indeed retrieve erroneous or partial patterns, which made it behave like a biological neural network!

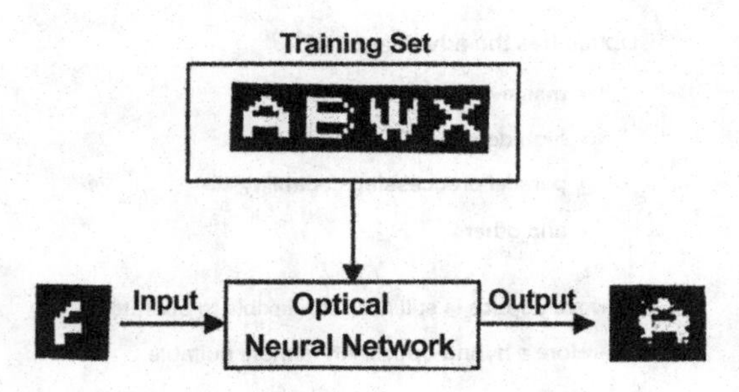

FIGURE A5 ONN Operation

Let us now move on to consider the neural networks that use the intra-pattern association algorithms. Intra-pattern association neural network means that the neural network ignores the association between the stored exemplars, while inter-pattern association neural network deals with association among stored patterns.

The differences between the inter-pattern association and the intra-pattern association are illustrated in Figure A6, where we see that twin brothers Tony and George look alike, except Tony has hair and George has no hair but does have a mustache.

In this figure we see hair and mustache used as special features that differentiate them. All other facial features they have in common. Thus we can see that by considering the special and common features, an inter-pattern association neural network can be developed to differentiate Tony and George using a simple logic operation algorithm.

Let us now show the performances, as compared between inter-pattern association and intra-pattern association neural networks, as demonstrated in Figure A7. In this slide we show the performance for inter-pattern association neural network on the upper part of the slide, while the performance for the intra-pattern association is in the lower part.

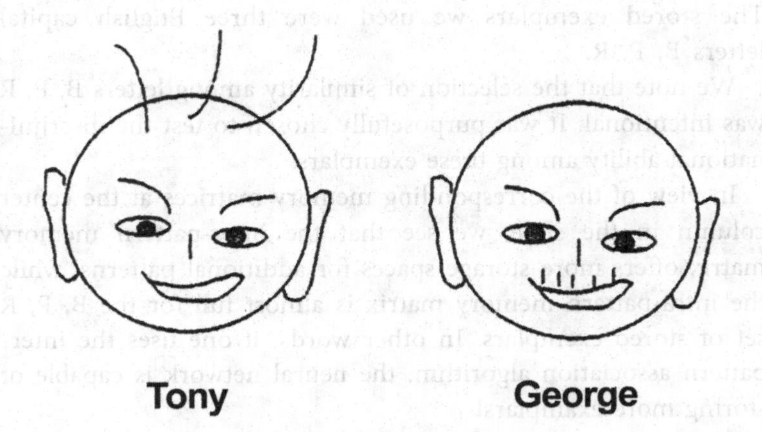

FIGURE A6 What is Inter-Pattern Association

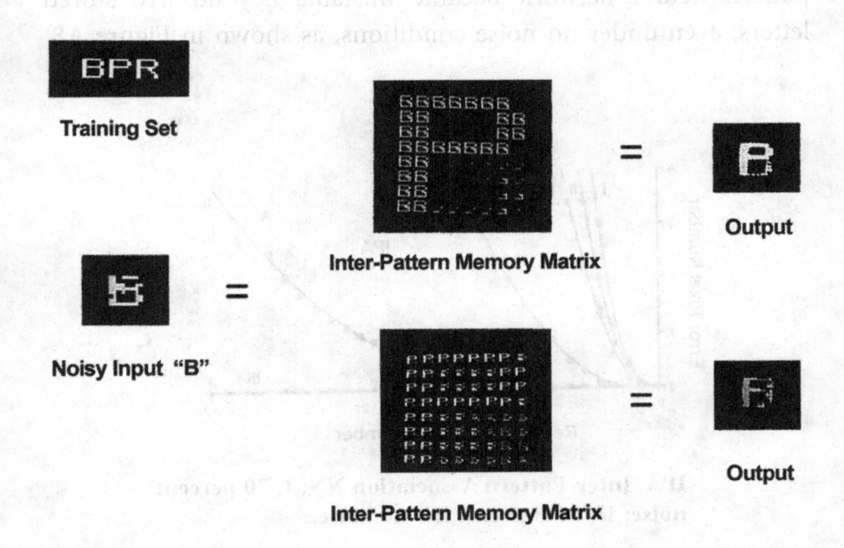

FIGURE A7 Inter-Pattern and Intra-Pattern Neural Networks

The stored exemplars we used were three English capital letters B, P, R.

We note that the selection of similarity among letters B, P, R was intentional. It was purposefully chosen to test the discriminational ability among these exemplars.

In view of the corresponding memory matrices at the center column in the slide, we see that the inter-pattern memory matrix offers more storage spaces for additional patterns, while the intra-pattern memory matrix is almost full for the B, P, R set of stored exemplars. In other words, if one uses the inter-pattern association algorithm, the neural network is capable of storing more examplars!

We further extended our training set to all twenty-six capital letters for the test in our 64-neuron optical neural network. In doing so, we were able to show that neural networks perform better for the inter-pattern association algorithm, for which the network can actually retrieve all twenty-six capital letters without error, under no input noise conditions. Whereas, the intra-pattern neural network became unstable beyond five stored letters, even under no-noise conditions, as shown in Figure A8.

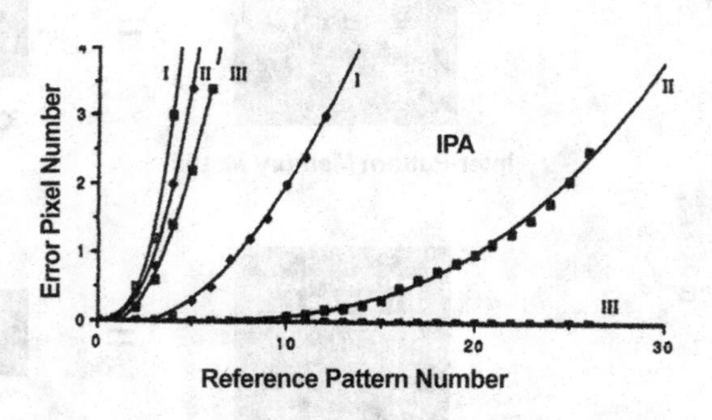

IPA, Inter-Pattern Association NN; I, 70 percent noise; II, 30 Percent; III, no noise.

FIGURE A8 Performance of Inter and Intra Pattern Neural Networks

Once again, we have shown that the artificial neural network (ANN) behaves like a biological neural network (BNN). Therefore it is clear that if one uses a smarter association memory one is able to store *more* patterns (remember more). In fact, those who use smart associative memory find it more fun to learn!

Note that the human brain is not designed to remember a huge quantity of information, although we have billions and billions of neurons! Otherwise, we would not have developed such storage gadgets as computers, compact discs, flash memory sticks, and other devices to help us.

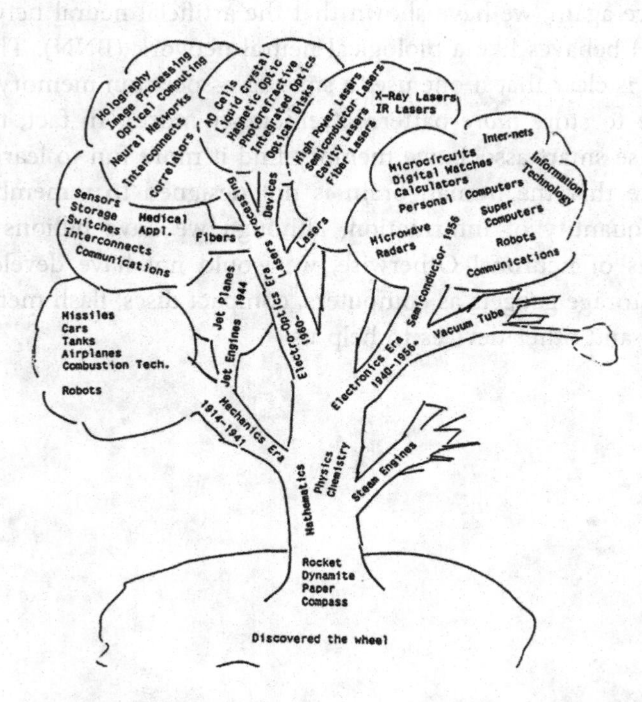

FIGURE A9 Technological Tree: Drawn by Yu for Students, 1987

Bibliography

N. Doidge. *The Brain that Changes Itself.* Penguin Group, 2007.

N. H. Farhat, D. Psaltis, A. Prata, and E. Pack. "Optical Implementation of the Hopfield Model." *App. Opt.*, 24, 1985, 1469.

J. J. Hopfield. "Neural Network and Physical System with Emergent Collective Computational Abilities." *Proc. Natl. Acad. Sci. U. S. A.*, 79, 1882, 2554.

T. Kohonen. *Self-Organization and Associative Memory.* Springer-Verlag, 1984.

D. E. Rumelhart and D. Zipser, "Feature Discovery by Competitive Learning" in *Parallel Distributed Processing: Explorations in the microstructure of cognition*, Vol. 1, D. E. Rumeihart and J. L. MaClelland, eds. Cambridge, Mass: MIT Press, 1988, 151–193.

L. K. Silverman, *Upside-Down Brilliance: The Visual-Spatial Learner.* DeLeon Publishing, 2002.

F. T. S. Yu and S. Jutamulia, *Optical Signal Processing, Computing, and Neural networks*, Wiley-Interscience, 1992.

About the Authors

Francis T. S. Yu received his B.S.E.E. degree from Mapua Institute of Technology Manila, Philippines, and his M.S. and Ph.D. degrees in Electrical Engineering from the University of Michigan. He has been a consultant to several industrial and governmental laboratories. He is an active researcher in the fields of optical signal processing, holography, information optics, optical computing, neural networks, photorefractive optics, fiber sensors, and photonic devices. He has published over five hundred papers, in which over three hundred are refereed. He is a recipient of the 1983 Faculty Scholar Medal for Outstanding Achievement in Physical Sciences and Engineering, a recipient of the 1984 Outstanding Researcher in the College of Engineering, was named Evan Pugh Professor of Electrical Engineering in 1985 at Penn State, a recipient of the 1993 Premier Research Award from the Penn State Engineering Society, was named Honorary Professor in Nankai University in 1995, is the corecipient of the 1998 IEEE Donald G. Fink Prize Paper Award, was named Honorary Professor of the National Chiao-Tung University in Taiwan, and is the recipient of the SPIE 2004 Dennis Garbor Award. He has served as an associate editor editorial board member, and a guest editor for various international journals. He is the author and coauthor of nine books. Dr. Yu is a life fellow of IEEE and a fellow of OSA, SPIE, and PSC. He was the recipient of the 2016 OSA Emmett Leith Medal.

Edward H. Yu is a teacher, writer, and the author of *The Art of Slowing Down: A Sense-Able Approach to Running Faster* (Panenthea Press) and *The Mass Psychology of Fittism: Fitness, Evolution & the First Two Laws of Thermodynamics* (Undocumented Worker Press).

Ann G. Yu lives and works in Los Angeles, California.

Printed in the United States
by Baker & Taylor Publisher Services

Printed in the United States
by Baker & Taylor Publisher Services